PARADIGMS REGAINED

PARADIGMS

REGAINED

A Further Exploration of the

Mysteries of Modern Science

John L. Casti

LITTLE, BROWN AND COMPANY

A *Little, Brown* Book

First published in the United States of America by
William Morrow and Company, Inc., 2000
First published in Great Britain by Little, Brown and Company, 2000

Copyright © John L. Casti, 2000

The moral right of the author has been asserted.

A CIP catalogue record for this book
is available from the British Library.

ISBN 0 316 64816 7

Book design by Richard Oriolo

Printed and bound in Great Britain by Clays Ltd, St Ives plc

Little, Brown and Company (UK)
Brettenham House
Lancaster Place
London WC2E 7EN

To the memory of Vivien M. Casti

Preface

The most distinguishing characteristic of science separating it from other reality-generating mechanisms like religion, mysticism, poetry, or music is that it is always tentative. Scientific theories represent our best guess at the moment as to how the world operates. But those guesses are continually being revised in the light of new observational evidence. For this reason there is nothing akin to the Ten Commandments in science; theories are created to be replaced.

In 1989, I published the volume *Paradigms Lost,* which aimed to give a layman's account of six of the great mysteries of modern science: the origin of life, the genetic determination of human behavioral traits, the acquisition of language, the creation of a thinking machine, the existence of intelligent extraterrestrial life, and the strange nature of quantum reality. In that book, I examined the competing positions in the scientific community on these fascinating questions; I discussed who held what view, and how they came to that position. The material presented there represented the state-of-the-scientific-art circa 1987–88, the time when I was actually writing the book. But a lot has changed in the succeeding decade or so. And as we enter a new millennium, the time seems ripe to revisit each of those stories and see where science stands today on the Big Questions. The volume you hold in your hands is that fast-forward journey through the last decade of advances on these issues.

To make the stories flow more smoothly, as well as to attach a kind of game-playing aspect to the account of the various theories, I structured each chapter in *Paradigms Lost* as a jury trial, in which the experts came in to give their evidence for or against the cases presented for the Prosecution and the Defense. At the end of each chapter, I stepped in to play the role of a member of the jury, giving my opinion as to which of the competing theories seemed to have made the strongest case. In the current volume I have maintained this courtroom fiction, presenting each Big Question in the form of an appeal to a higher court of the verdict reached in *Paradigms Lost*. Again, I invite the reader to examine the evidence and come to his or her own conclusion as to whether the earlier verdict should be set aside or let stand.

In order to make the current volume self-contained, I have presented a short summary of the competing positions given in *Paradigms Lost* at the beginning of each chapter. This is a highly compressed overview of much more detailed arguments, for which I refer the interested reader to *Paradigms Lost* for a full account.

Let me take this opportunity to express my gratitude to Rebecca Goodhart, Katharine Cluverius, and Stephen Power, editors extraordinaire, who insisted that I write these stories so that they could understand them. Everyone, especially the author, has and will continue to benefit from their good deeds.

—JLC
SANTA FE and VIENNA

Contents

PARADIGMS REGAINED

ONE

The Truth, the Whole Truth, and

the Scientific Truth

The State of Science Today

The Why of the World

Why do I see what I do and do not see something else? This question forms the basis for a large fraction of human inquiry into the universe around us. In a vastly oversimplified sense, the answer to the question is easy: We see what we do and do not see something else because of the *way* in which we look. And these "ways" constitute what I often term *reality-generating mechanisms*. Religion, mysticism, poetry, music, literature, and art are all such mechanisms. So is science. And each of these reality-generating schemes has its own characteristic set of tools and methods for answering the question. The methods of science consist of explanation of observed natural and human phenomena by producing a set of rules. I'll say more about these rules later. But for now it suffices to say that these rules are of a special type and, in contrast to many other reality-generating procedures, are always subject to revision in the light of new evidence. The examination of new evidence and its evaluation is what the stories in this book are about. Thus science, unlike religion, for example, is always tentative, which is the most important take-home message that the reader can get from this book.

Science has come to occupy a role as the final arbiter of many questions puzzling people in everyday life. This fact lends an air of credibility and respect to scientific pronouncements that most reality-generating mechanisms aspire to—but seldom achieve. As a result, a whole universe of practitioners has emerged whose statements and methods give the surface appearance of answering basic questions by appeals to "science." Such *pseudoscience* is the bane of every serious scientific researcher. As we go along through the stories of this volume, the reader should continually be aware of the distinction between true science and such superficial appeals to the words of science, and attempt to separate the words from the deeds. As an aid in this direction, let me recount a delicious hoax played on the high priests and priestesses of literary criticism and the sociology of thought a few years ago. The reader can use this example as a yardstick with which to compare pseudoscience with "the real thing," as we make our way through the remainder of the book.

Down Among the Sociologists

Alan Sokal is a mathematical physicist with a sense of humor. So in late 1994, he submitted a sham article titled "Transgressing the Boundaries—Toward a Transformative Hermeneutics of Quantum Gravity" to the cultural studies journal *Social Text,* in which he reviewed various current topics in physics and mathematics. Tongue in cheek, Sokal drew various cultural, philosophical, and political morals that he felt would appeal to fashionable academic commentators who question the claims of scientific objectivity.

The editors of *Social Text* didn't detect that Sokal's article was a hoax, and they published it in the Spring/Summer 1996 issue. Sokal himself revealed the hoax in an article for another journal, *Lingua Franca,* in which he explained that his *Social Text* article had been "liberally salted with nonsense," and in his opinion was accepted only because it sounded good and flattered the editors' ideological preconceptions. The main value of Sokal's hoax is that it served to attract attention to what many see as a decline of standards of rigor in the academic community. To fix the thrust and style of Sokal's text, here are a couple of extracts:

> Differential topology has traditionally privileged the study of what are known technically as "manifolds without boundary." However, in the past decade, under the impetus of the feminist critique, some mathematicians have given renewed attention to the theory of "manifolds with boundary."

> However, these criteria [that unobservable quantities should not be introduced into a scientific theory], admirable as they are, are insufficient for a *liberatory* postmodern science: they liberate human beings from the tyranny of "absolute truth" and "objective reality," but not necessarily from the tyranny of other human beings.

There are many targets of Sokal's satire, cutting a broad swath across the intellectual landscape. These include:

• *Postmoderns*—those in the humanities who like to surf through avant-garde fields like quantum field theory or chaos theory to dress up their own arguments about the fragmentary and random nature of experience.

• *Deconstructionists*—the sociologists, historians, and philosophers who see the laws of nature as social constructions.

• *Cultural critics*—those who find the whiff of sexism, racism, colonialism, militarism, or capitalism not only in the practice of scientific research but even in its conclusions.

By way of comparison with Sokal, and to give just a taste of the kind of gobbledygook regarding science that practitioners of the foregoing arts are capable of, here is a quote from that oracle of deconstruction Jacques Derrida:

> The Einsteinian constant is not a constant, is not a center. It is the very
> concept of variability—it is, finally, the concept of the game. In other
> words, it is not the concept of some*thing*—of a center starting from
> which an observer could master the field—but the very concept of the
> game.

What could this possibly mean? One can only hope that such incompetents
who pontificate about science as a social phenomenon without understanding
the first thing about its content are on the way out, and will soon be as rare as
blind art critics.

Sokal's revelation of his hoax drew an angry response that he had
abused the trust of the editors of *Social Text* by using his credentials as a
physicist. One of the editors even speculated that Sokal's parody was nothing
of the sort, and that Sokal had had a change of heart about his own views of
science! Perhaps the last word in what the results of physics imply about culture, politics, or philosophy should go to Steven Weinberg, Nobel laureate in
physics and commentator on the whole *affaire Sokal,* who stated, "The discoveries of physics may become relevant to philosophy and culture when we
learn the origin of the universe or the final laws of nature, but not for the present."

This Sokal affair shows that using the scientific terms without their content is like using soap without water—it doesn't clean up the situation but only
further muddies it. The end result is the spreading of confusion and disillu-
sion with science. Perhaps all we can really do is take solace in Nietzsche's
poignant observation, "Against ignorance the gods themselves struggle in
vain." Actually, the situation is even just a bit worse. Here's why.

The View from Edinburgh

Andrew Pickering, a sociologist at the University of Illinois, has stated,
"There is no obligation upon anyone framing a view of the world to take
account of what 20th-century science has to say." This outrageous statement
is symptomatic of a group of philosophers, sociologists, and historians known
as the Edinburgh school. They contend that knowledge produced by the natural sciences is a cultural construct essentially equivalent in its attributes to

the knowledge produced by less tractable reality-generating mechanisms, in which reproducible phenomena under controlled conditions either do not exist or are of little interest. Let's take a page or two to examine this extraordinary claim.

The Edinburgh school doesn't see itself as opposing science, or even questioning the integrity of scientists. Rather, it contends that scientific knowledge is only a communal belief system with a dubious grip on reality. Pickering's book *Constructing Quarks,* which analyzes the birth of the Standard Model of particle physics, is perhaps the most instructive, ambitious Edinburgh study of modern physics. In this volume, he focuses on cultural factors of which scientists are often unaware, and provides many solid insights into their significance. His conclusion, though, is eye-opening:

> The quark-gauge theory picture of elementary particles should be seen as a culturally specific product . . . a communally congenial representation of reality. Given its cultural resources, only singular incompetence could have prevented the high-energy physics community producing an understandable version of reality at any point in its history. . . . The preponderance of mathematics in particle physicists' account of reality is no more hard to understand than the fondness of ethnic groups for their native language.

And just how did Pickering come to this astounding conclusion? Fundamentally, this conclusion and others from the Edinburgh school emerged because the program's methodology is applied by them to scientific knowledge itself, and not just to practice.

Pickering's statement about the irrelevance of twentieth-century science stems partly from a recurring misunderstanding in sociological studies that what scientists see as progress often entails abandoning familiar but perplexing phenomena. In the case of the development of the Standard Model, the commonplace phenomena of an empirical flavor were abandoned in favor of more esoteric phenomena that could only be characterized mathematically. But, in fact, the old phenomena were not really abandoned but simply set aside, just as Galileo set aside friction. The "new" particle physics was born of the recognition that the "old" phenomenology was less acceptable than that on which the new physics focused.

The Edinburgh sociologists have objectives resembling those of the philosopher of science Thomas Kuhn, who was extensively discussed in the opening chapter of *Paradigms Lost.* That sociological factors have been important in the development of scientific knowledge was amply demonstrated by Kuhn, who is recognized as a forerunner by the Edinburgh school. Both see claims for scientific advances as largely circular arguments because they don't conform to the mythical format in which observation always precedes theory, with logic and solid data always pointing to one, and only one, theory. This forces them to replace "always" by "never," which is yet another oversimplification because the correct word to use here is "sometimes."

Many of the misconceptions in the sociology of science literature follow from the methodology of the program of the Edinburgh group. For example, the requirement that rationality and irrationality be treated symmetrically gives equal credence to contending positions of objectively unequal merit, such as whether or not cold fusion exists. This leaves only psychoanalysis or social construction as explanations for the scientists' choice. And predictive power, the strongest evidence that the natural sciences have an objective grip on reality, is largely ignored by these researchers.

So we come to the conclusion that while science is manifestly a cultural phenomenon, facts like those just stated cannot be reconciled with Pickering's statement that "the science-as-knowledge image of science is so thin that it bears almost no relation to its subject matter." It does and will continue to do so for the foreseeable future.

But the sociologists are not the only ones who level their guns against science. Even scientists themselves do it when they take off their lab coats and mount the pulpit. From the commonsense view that "science cannot know everything" it's but a small step to the idea that religion may be a more satisfactory way of organizing one's idea of one's place in the world. Basically, what both the preachers and the sociologists are doing is forming a belief system, or what I prefer to call a *reality-generating mechanism.* As noted at the beginning of the chapter, such mechanisms are devoted to answering the question, Why do I see what I do, and do not see something else? Poetry, literature, art, mysticism, and music are other such reality-generating schemes. But the main competitors in this battle of the minds remain science and religion. So let's briefly revisit this theme, treated extensively in *Paradigms Lost,* with an eye toward seeing how scientists can possibly reconcile their religious beliefs with the tenets of their profession.

Science and God

In 1993, best-selling British author Susan Howatch contributed 1 million pounds to establish the Starbridge Lectureship in Theology and Natural Science at Cambridge University. Frequenters of airport bookshops will know of Howatch as the author of blockbusters such as *Penmarric, Wheel of Fortune,* and *Glittering Images,* a story of clerical life in the fictional British town of Starbridge, for which the lectureship is named. According to Howatch, "Science and theology are no longer seen as opposed, but complementary, two aspects of one truth."

Standing up for the traditional separation between science and religion, the prestigious British science weekly *Nature* stated in an editorial that the complementary aspects of science and religion

> can only be true in the most superficial sense, that in which some people see one truth and not the other, or vice versa. For the many people who take the scientific and the theological (or at least religious) view together, it is more common to reconcile the inevitable intellectual conflict on matters such as the after-life by supposing that there are two truths, not two aspects of one truth, or otherwise to suppose that Bible-stories or their equivalents in other religions have an allegorical function of great moral value to believers.

So what is one to make of this narrowing of the divide between the godless and the godly? Is there really a warming trend moving through the scientific community concerning its acceptance of the Deity? Can rational inquiry and spiritual conviction be reconciled? It's of interest in this regard to see the results of a survey of 1,000 randomly selected scientists done in 1916 and again in 1996 as to their belief in a personal God. The results are displayed in Table 1.1.

BELIEF IN A PERSONAL GOD	1916	1996
Personal belief	41.8%	39.3%
Personal disbelief	41.5%	45.3%
Doubt or agnosticism	16.7%	14.5%

Table 1.1. Survey answers in 1916 and 1996

What's striking about this result is the remarkable consistency over the eighty-year period between the two surveys despite all the discoveries Pickering would have us ignore. In these years, religious belief has become more diverse. But it seems that traditional Western theism has not lost its place among U.S. scientists. It's worthy of note, however, that despite the stability in the overall proportion of believers and disbelievers, there was a significant shift in views held by the three disciplines surveyed—mathematics, biology, and physics/astronomy. The 1996 survey showed that mathematicians are the most inclined to believe in God (44.6 percent). And although biologists showed the highest rates of disbelief in 1916 (69.5 percent), that ranking now goes to the physicists and astronomers (77.9 percent). As Leon Lederman, Nobel Prize winner in physics in 1988, says, "Science has turned up no proof of the divine, and although at the edges of science there is the unknown, and that leaves room for a creator, there is a lot less room than 50 years ago. The space available for God seems to be shrinking."

By way of rebuttal, John Polkinghorne, president of Queen's College at Cambridge University, a physicist for twenty-five years before becoming an Anglican priest, notes that "the trend is to look for God in dramatic discontinuities in physics or biology, and if none are found, to declare religion vanquished. But God may act in subtle ways that are hidden from physical science." Christian de Duve, another Nobel Prize winner (in 1974 for his work on molecular biology), says, "Many of my scientist friends are violently atheist, but there is no sense in which atheism is enforced or established by science. Disbelief is just one of many possible personal views."

The biologist Richard Dawkins, well-known for his many popular books on evolution such as *The Selfish Gene, The Blind Watchmaker,* and *River out of Eden,* states:

The universe we observe has precisely the properties we should expect if there is at bottom no design, no purpose, no evil, and no good, nothing but pointless indifference. . . . In a universe of selfish genes, blind physical forces and genetic replication, some people are going to get hurt, others are going to get lucky, and you won't find any rhyme or reason for it.

These words echo the famous closing sentence in Nobel laureate Steven Weinberg's best-selling book *The First Three Minutes,* where he stated, "The more the universe seems comprehensible, the more it also seems pointless." In a later interview, Weinberg adds, "What we are learning about physical law seems coldly impersonal and gives no hint of meaning or purpose."

So there we have it. The believers, the nonbelievers, and the I-just-don't-knows. What does this really tell us about science and scientific truth? In the final analysis, probably not much beyond the fact that science is still practiced by human beings, beings who are quite capable of holding mutually contradictory views on a whole host of questions. The faith-based view of religion versus the rationality of the scientific method is just one of those sets of contradictions. On this less than conclusive note, let me try to give some balance to this introductory section by taking a look not at the misuse and perversion of science, but at the way science actually does work.

The Scientific Scheme of Things

Write down the sequence 0, 3, 6, 12, 24, and so on, where each number is obtained by doubling its predecessor. Now add 4 to each number and divide the result by 10. This produces the sequence 0.4, 0.7, 1, 1.6, 2.8 . . . If you now take any astronomy book and look up the distances of the planets from the Sun, taking the Sun-Earth distance as 1, the distances are virtually identical to the terms in this sequence for all but the outermost planets.

This relationship has puzzled astronomers ever since it was first published in 1766 by the astronomer Johann Daniel Titius. One camp clings to the belief that this "law" is just a coincidence, while others feel that it reveals a hitherto undiscovered feature of the solar system.

For a mere coincidence, the relationship has an astonishing track record. Shortly after its publication, Titius's compatriot Johann Bode suggested that the lack of a known planet at 2.8 units cries out for a search for the "missing" planet between Mars and Jupiter. By 1796, Bode had convinced astronomers to hunt for it, and in 1801 it turned up: The minor planet Ceres, just one thousand kilometers across was found orbiting the Sun at exactly 2.8 units. Earlier the Titius-Bode rule placed Uranus at 19.2 units, astonishingly close to what turned out to be its actual 19 units when the planet was discovered in 1781. But the "law" ran into trouble with Neptune and especially Pluto. The latter is at about 40 units, a far cry from the 77.2 units predicted by the Titius-Bode rule.

So were all the earlier successes just flukes? Today, conventional wisdom in the astronomical community sees the rule as simply a numerological curiosity. But recently two French astronomers, François Graner and Bérengére Dubrulle, have found evidence to suggest that the rule is a natural consequence of certain symmetry properties that are almost certain to feature in any planetary system. So perhaps the Titius-Bode law may reflect some deep significance about the solar system, after all.

Put simply, the two French astronomers discovered that all models for the formation of planetary systems have two symmetries: *rotational invariance* and *scale invariance*. The first means that no matter how the cloud of material, which coalesces around the central star to form the planets, is turned, it always looks the same. The second invariance means that the cloud and its contents look the same on all scales of length. After a bit of mathematical wizardry, what comes from these observations is a slightly modified form of the Titius-Bode law in which instead of doubling each term to form the initial sequence of numbers, we multiply by 1.7, and then raise the result to the first power for Mercury, the second for Venus, and so on, finally multiplying each result by 0.23.

It's still an open question whether the two symmetries leading to this extended Titius-Bode law actually prevailed in the early days of the real solar system. But the process by which this new version of the law was arrived at is an admirable example of the workings of the scientific enterprise.

Relations, Laws, and Theories

The Titius-Bode law illustrates not only the methodology of science, but the all-important fact that science, unlike many of its competitors in the reality-generation game like religion, is not in the business of providing absolute answers to questions about the world around us. Science offers *tentative* theories, which are always subject to modification in the light of new evidence. And how are these theories created? Fundamentally, the procedure is what has come to be known as the *scientific method*. It involves the steps contained in the following diagram:

$$\text{Observations/Facts}$$
$$\downarrow$$
$$\text{Hypothesis}$$
$$\downarrow$$
$$\text{Experiment}$$
$$\downarrow$$
$$\text{Laws}$$
$$\downarrow$$
$$\text{Theory}$$

In this method, observations, like the planetary positions, give rise to conjectures and hypotheses like the Titius-Bode law, which in turn are tested by performing experiments. If the experiments don't confirm the hypothesis, a new hypothesis is formed, just as in the work described above by Graner and Dubrulle. Those hypotheses that survive are encapsulated into empirical relationships, or laws, like the Titius-Bode law, which in turn are embedded in larger explanatory theories (in this case, Newton's theory of planetary motion). It is this sequence of steps that has been the focus of most of the philosophical analyses of the methodology of science.

Of course, there's more to the scientific enterprise than just concocting a theory. Not just any old theory will be accepted by the community of scientists as a good theory of something like the positions of the planets. Candidate theories must be vetted by imposing additional criteria attesting to their quality. Foremost among these criteria are the verifiability of claims and peer review.

• *Verifiability of claims.* Science is a public undertaking with many filters that a claim must pass through before it's accepted as part of the current conventional wisdom. Two of the most important of those filters are the refereeing process for scientific articles and the repeatability test for experimental results. Before a reputable scientific journal will publish a research announcement, it is sent out for review to other workers in the field, not only as insurance that the results are correct, but also to substantiate their significance with respect to the current state of knowledge in the area. In a similar manner, published work is supposed to report all the details of the investigator's experimental setup so that any interested party can, in principle, repeat the experiment and try to replicate the reported results. Thus, in the utopian world where the scientific ideology reigns, refereeing and replicability of results keep the scientific process (and the scientist) honest.

• *Peer review.* The peer review process involves committees of experts from the various scientific fields getting together and recommending to the funding agencies and editors of journals those projects and those scholars whose work they feel merits support. According to the peer review rationale, this process ensures that money and public recognition of ideas are channeled to those institutions and individuals showing the clearest evidence of being able to do something productive with this support.

Given the conventional view of the scientific enterprise as highly rational and objective, it comes as no surprise that many scientists accept this as at least a very close approximation to the way all scientists really are, possessing an impartial, egalitarian, and meritocratic nature. However, this conventional image focuses entirely upon the *process* of science, leaving aside all consideration of the motives and needs of the scientists themselves. The degree to which this omission casts a cloud over the rosy picture painted above will be a subtheme throughout much of this book. For now, I'd like to draw the reader's attention to one of the most interesting phenomena affecting the public's general view of the scientific enterprise, what one might term the *mediazation* of science. It can undermine the peer review process—or bypass it altogether. A good starting point is how the fickle finger of fate came down and touched the life of the Cambridge mathematical physicist Stephen W. Hawking.

A Brief History of Hawking

In summer 1992, Stephen Hawking's little volume, *A Brief History of Time,* became the all-time British best-seller, with its record-breaking number of

consecutive weeks on the best-seller list. It was certainly the publishing sensa-
tion of the late 1980s and early '90s and a deeply mysterious one. Survey after
survey turned up the not very surprising fact that hardly anyone who bought
the book actually read it, and hardly anyone who read it understood it, plung-
ing as it does into oceanic abstractions of space-time singularities, charmed
particles, superstring theory, and the finite universe that has no beginning and
no end. In short, nobody but a trained physicist could understand half of it,
and the trained physicists all say they already knew everything the book had
to say anyway. So why did people buy it?

Well, incomprehensibility itself is probably a selling point. Like the
Bible in the bookcase in another more leisurely, bygone era, you may not
understand Hawking's book but at least you have the feeling that *somebody*
understands it, and that by having the book you possess that understanding
even if you are not clever enough to share it.

Another major factor contributing to the book's success is the image of
Hawking himself: the lonely, indomitable genius, crippled since the age of
twenty-three with Lou Gehrig's disease, an affliction that was expected to kill
him within two years, but has left him with control of two fingers that he uses
to operate his wheelchair and the synthesizer that generates his unworldly
voice. It seems people like a genius to be handicapped. Witness Beethoven's
deafness, Milton's blindness, or Byron's gamy leg.

A third factor in the book's success is the title, which is catchy, paradox-
ical, and very promising. A history of *time*. What could be more alluring? The
Hawking phenomenon itself is but a small slice of a much larger picture: the
scientist—and science, itself—as a media personality. A little more amplifica-
tion of this theme may help shed some light on the way certain theories and
theorists tend to get their ideas taken over by the public as the received wisdom of science.

In from the Cold

On March 23, 1989, two noted chemists called a press conference at the University of Utah in Salt Lake City. The announcement they made was stagger-
ing. The scientists, Martin Fleischmann and B. Stanley Pons, claimed to have
produced controlled nuclear fusion at room temperature in a test tube.

Front pages of newspapers around the world, as well as the covers of *Time* and *Newsweek* magazines, immediately brought the news of this "discovery" to the public. The result of a relatively simple experiment could soon provide the world with safe, clean, and dependable energy—the dream of humankind for centuries.

Ironically, only thirty-five miles away from Salt Lake City, Steven Jones of Brigham Young University in Provo, Utah, had been experimenting with cold fusion but was finding that it could not produce energy in sufficiently large amounts for practical use. Jones had earlier learned of the Fleischmann and Pons claims and had, he thought, reached an agreement with the University of Utah scientists to release both sets of results simultaneously. The precipitous announcement on March 23 ignited controversy not only over whose experiments were correct, but also over whether nuclear fusion was actually occurring or if the results were due to other, possibly unknown, effects.

A decade later, the cold fusion story has been pretty much played out. The balance of evidence seems to indicate that there is no such thing, at least not with the kind of tabletop apparatus claimed by the Utah researchers. But even to this day the controversy rages on in some limited circles, with evidence being cited by some of cold fusion occurring in laboratories and, possibly, even in the mantle of the earth. However, that's not what is of particular interest for us in this book. For us, it is the sociological aspects of the cold fusion affair that provides a lesson in how science, like many other aspects of modern life, has been packaged, sliced, diced, and digested for public consumption by the media.

To many in the scientific community, the strangest aspect of the Fleischmann-Pons announcement of their cold fusion claim was that it was done not in the pages of an august periodical like *Nature* or *Science,* but at a press conference! Granted that everyone agreed that the results, if true, were, as one scientist noted later, "as important as the discovery of fire"; but to announce any type of scientific result, let alone one of this magnitude, at a press conference before peer review of that result was completely unprecedented—and highly unprofessional.

The obvious conclusion to draw from this way of announcing their results is that Fleischmann and Pons decided to roll the dice and get as much leverage out of their work as possible, before the flinty-eyed reviewers started tearing into their very shaky experimental procedures and analysis. And, in fact, this strategy worked quite well—for a while. The state of Utah immediately produced a ton of money for a cold fusion institute at the university, the scientists themselves became fifteen-minute media darlings, and cold fusion was on everyone's lips. Alas, it came to no good end. The cold fusion affair is a textbook example of science-by-media rather than by peer review and experiment. And it's not an isolated singularity. It's by now difficult to count the number of similar cases of *mediazation* of science in the last decade—life on Mars, the cloning of Dolly, Fermat's Last Theorem, all represent cases of the media jumping the gun before the science is even close to being settled. The main lesson to be learned from all this is simple: Don't believe everything you read in *The New York Times*.

But what can we make of the celebrity status into which Hawking's impenetrable volume catapulted him? In some ways this is even more interesting than the sad case of Pons and Fleischmann, since it focuses on the scientist more than on the science. By common consensus in the theoretical physics community, it is certainly not Hawking's work that accounts for his public appeal. John Barrow, professor of astronomy at the University of Sussex, notes, "To compare Hawking to Newton or Einstein is just nonsense. . . . In a list of the 12 best theoretical physicists this century Steve would be nowhere near." So the reason must lie elsewhere, most likely in the type of legends that the public creates, often through literary depictions, of scientists. So let's have a look at six such stereotypes identified by Roslynn Haynes of the University of New South Wales in Australia, by which the literary community has represented the scientist. Perhaps one of them will fit Hawking and give some insight into why the public, if not his peers, regard Hawking with such awe.

Portraits of the Scientist in Words

The old adage "Truth is stranger than fiction" never applied more forcefully than to the fictional portrayal of scientists. Through centuries of literature, writers have recycled six standard stereotypes, each with its own myths, within which to slot scientists and their projects. Let's end this brief introductory chapter with a summary of these six stereotypes, which readers can use as a starting point in assessing their own images of scientists and how scientific theories are created.

• *The evil alchemist:* Alchemists have always been regarded as sinister magicians, usually in league with the devil. In literature, the alchemist has been represented as engaged in illegal and even sinful research, working in secret, but also as a proud and rebellious individual. The archetypal alchemist was Dr. Faustus, whose hubris led to eternal damnation. Faust and his successor, Dr. Frankenstein, have provided the most consistent and ready parody of legitimate scientists engaged in cutting-edge research in three particular fields: physics, biology, and medicine. The prototypical representation depicts these alchemist/scientists as arrogant, power-crazed, secretive, and insane in their pretensions to transcend the human condition. Scientists are still perceived by many laymen as powerful, frightening, and isolated figures, speaking a language and thinking thoughts accessible only to their colleagues.

• *The noble scientist:* Sir Francis Bacon's *New Atlantis,* published in 1626, was the first literary work to portray the scientist in a positive light. In this work, Bacon attempted to revolutionize the image of the scholar from one of a mercenary pedant to that of an altruistic idealist, intent only on contributing to the common good. *New Atlantis* provided the motivation for the founding of the Royal Society of London in 1662. The archetype of the noble scientist was Newton: humble where Faust had been proud, seeking to explain rather than to mystify, and to bring harmony and order to the heretofore confusing cosmos. Since World War II, the image of the noble scientist has ceased to appeal and such stereotypes are seldom invoked, except as a yardstick against which to measure the deficiencies of contemporary scientists.

• *The stupid obsessive:* In Thomas Shadwell's 1676 play *The Virtuoso,* Sir Nicholas Gimcrack flits from one trivial curiosity to another, without rigor or logic, duped by the purveyors of expensive equipment and fake "wonders" and supremely uninterested in either the study of mankind or family responsibilities. Gimcrack provides a stereotype that has appeared rather frequently since. For example, in Gulliver's travels to the island of Laputa, he tours the Grand Academy of Laputa, where he encounters scientists empowered to make decisions merely on the basis of their shortsighted specialist knowledge of one factor out of many in a complex situation. The absent-minded professors of comic strips and films are attenuated versions of Gimcrack, so engrossed with their research that they wear nonmatching socks, never cut their hair, and remain oblivious to the dangers confronting their beautiful daughter in the next room. Albert Einstein is a leading example of just this sort of stereotype.

• *The inhuman rationalist:* The scientific materialism of the Enlightenment generated perhaps the most enduring scientist stereotype, that of the inhuman researcher who has sacrificed the humane, emotional side of himself to the purely rational and intellectual, abandoning all human relationships in an obsessive pursuit of science. Nineteenth-century literature abounds with such characters who have rejected human bonds for the pursuit of scientific discoveries, the most famous being Mary Shelley's *Frankenstein.* Victor Frankenstein epitomizes the Romantic anathema, the man, who in pursuit of science, rejects father, fiancée, nature, and even his surrogate child, the Monster. In our century, the disciplines of physics, mathematics, and computer science have provided the foremost examples of the inhuman rationalist. Scientists working on the Manhattan Project and afterward during the cold war have especially fueled this stereotype with their declarations of unconcern about the human costs of their inventions.

• *The adventurer hero:* One of the most attractive, albeit simplistic, scientist stereotypes emerged in the late nineteenth century when belief in progress and realization of the commercial results of technology cast science as an obedient servant, an empowering tool. The literary representative of this change of view was the scientist-adventurer, the counterpart of the Romantic hero. Perhaps the first expression of this stereotype was in the novels of Jules Verne. His myths of conquest present debonair scientist heroes defeating the marvels of nature with marvels of science. Arthur Conan Doyle's Professor Challenger stories are cut from the same mold, purveying a potent mix of science, adventure, courage, and moral superiority. In our time, Verne's and Doyle's adventurers have become space travelers, still representing the might-is-right ethic. The heroes of the ever-popular television series *Star Trek* provide an illustration of this stereotype.

• *Out of control:* The theme of the scientist hoist with his own petard has always been a popular one in antirationalist times. *Frankenstein,* for example, focused on the scientist's refusal to foresee or accept responsibility for the disastrous results of his research. Nowadays rationalism is again under attack and antiscientific ideologies have a large following. So, scarcely a project backfires without the media resorting to this convenient shorthand. Nuclear power, genetic engineering, in vitro fertilization, organ transplantation, industrial pollution, the human genome project, virtual reality, are among the many projects popularly presented as examples of science out of control. The modern heirs of Dr. Frankenstein are depicted as being, at best, ignorant of the likely ecological and sociological implications of their research and, at worst, liable to suppress such a realization lest ethics committees get in the way of their grants.

The single most characteristic feature of science is that its conclusions are always tentative, ready to be overthrown by new observational evidence and new theories that more compactly, more elegantly, and/or more completely explain that evidence. On the other hand, the public's perception of science—and scientists—is often completely at variance with this always-ready-to-change spirit, seeing science as rigid, unyielding, cold, remote, and even hostile to human values and culture. Here I want to challenge the reader to consider the way science actually does work, as it proceeds to create new theories—and new realities—in the light of new observations about man and the universe coming from places as diverse as the Hubble space telescope and the *Deep Sea Challenger* minisubmarine. Solving the Great Mysteries of science is akin to peeling the skin of an onion. We get closer and closer to the core—but never quite reach it. Here our goal is simply to peel back one more layer, and give an account of where science stands today on the fascinating questions residing in that twilight zone where science becomes philosophical and philosophy is becoming scientific.

Paradigms Regained

Paradigms Lost is a presentation, made to the intelligent layperson, of six of the big unsolved problems of modern science: the origin of life, the genetic basis of human social behavior, the way humans acquire spoken language, the creation of a "thinking machine," the existence of intelligent extraterrestrial life, and the nature of quantum reality. *Paradigms Regained* reexamines all six of these questions, looking at the many developments that have occurred in the nearly twelve years since *Paradigms Lost* was written (1987–88).

In order to make the message a bit easier to swallow, I structured each chapter of *Paradigms Lost* as a jury trial. Each of the Big Questions was posed as a claim, with witnesses for the Prosecution and Defense marching to the stand to present their evidence. At the end of each chapter, I took the position of a member of the jury, stating how I would vote on the competing claims and inviting the reader to do the same. I've employed the same kind of fiction, of the jury trial, in this volume, presenting each chapter as an appeal to a higher court of the verdict reached in *Paradigms Lost*. Thus, the legal issue to be settled is whether the appeal for a new trial should be granted or if the previous verdict should be left standing.

The fine details of personalities, sideshows, and off-the-track discussions relating to each question can be found in *Paradigms Lost*. But in order to make this sequel entirely self-contained, I have presented a short summary of the arguments and the verdict reached in that earlier work at the beginning of each chapter of this book. Nevertheless, I urge the interested reader to consult the previous volume for a wealth of detail and references that I did not have space, interest, or need to repeat here.

Now, without further ado, let's go before the bench.

TWO

That's Life!

Claim: Life Arose out of Natural

Physical Processes Here on Earth

BACKGROUND

What Is Life?

In 1944, as the guns of World War II roared over Europe, the Dublin Institute for Advanced Studies served as the venue for the opening of a new era in science. There, the institute's director, Erwin Schrödinger, delivered a set of lectures that set the course for a whole new science. At the time, Schrödinger was already known as one of the architects of the quantum theory, for which he had been awarded the Nobel Prize for physics in 1933. But in the scope of history, it may well turn out that Schrödinger ends up being even more well known for this set of lectures, which ushered in what we now call the field of "molecular biology." These lectures, which were later published under the title *What Is Life?*, drew explicit attention to the vexing question, How do we decide whether some collection of matter possesses the ineffable quality of being "alive"? More precisely, what is it *exactly* that tells us that a stone on the street is not alive, while our family dog does have this quality? As a man living with his wife and mistress under the same roof in Roman Catholic Dublin in the 1940s, Schrödinger clearly had good reason to reflect on the nature of the human condition—including life. And his little book served as the springboard for the stunning discovery of the double helical structure of the DNA molecule by Francis Crick and James Watson in 1953, the event that really marks the public birth of modern biology.

When it comes to defining what it means to be alive, there are as many answers as there are biologists. For example, Carl Woese, an evolutionary biologist at the University of Illinois, says, "Life is simply this: an entity that can make a copy of itself from parts all of which are far simpler than itself." But even a self-replicating robot would be alive by this definition. And that doesn't have the right ring to it, since human engineers would have had to build the first one. So Woese corrects himself in mid-course, saying, "Evolutionary history is going to have to be part of any really workable definition of life." Thus, to be alive, an object must not only create itself from simpler parts, but must also participate in the grand design of evolution.

Since putting forth a definition of life is a game that absolutely everyone can—and does—play, let me cut to the chase and offer my own criteria for qualifying an entity as being alive. For our purposes in this chapter, an object is living if it displays the following three properties:

- *Metabolism:* The object takes resources from its environment, and then processes those resources in order to sustain its own existence.

- *Self-repair:* The object is capable of rectifying errors in either its metabolic or genetic machinery.

- *Replication:* The object is capable of fabricating good—but not necessarily perfect—copies of itself.

Okay, that's it. Just those three properties—metabolism, self-repair, and replication. If you've got them, then you're alive; otherwise, you're not.

The most evident feature of modern life is that the metabolic and replication processes are completely intertwined. In order for a living cell to carry on its business of replication, the genetic machinery needs the help of enzymes (proteins) that are produced by the cellular metabolic machinery. Similarly, without the genetic information the metabolic activity could not be carried on. This is the so-called "gene-protein" linkup, which is common to all life-forms on Earth today. But it was not always this way. So, one of the biggest challenges that any viable candidate for a theory of the origin of life must face is to explain first how either the replicators or the metabolizers came about and were able to carry on their activity without the other. Then the theory must give a plausible account of how the gene-protein linkup was established.

Competing answers to these foundational questions come in three essentially different flavors.

1. *Earthly theories:* These theories all posit that the first living form came about through natural physicochemical processes taking place on Earth around 4 billion years ago.

2. *Off-Earth theories:* Theories in this category all claim that one way or another, the first living object was transported to Earth from outer space.

3. *Mystical theories:* All theories in this category invoke a supreme deity, who is responsible for creating life de novo. These theories are essentially nonscientific. But as they are very prevalent in some circles, we will consider them briefly later in the chapter.

Since a variety of theories in each of these categories were examined in detail in Chapter 2 of *Paradigms Lost,* here I will focus most of my attention on the new theories not discussed there, saying just a few words about the competition from a decade ago. The remainder of the chapter will address new developments pertaining to the earlier theories, as well as an account of how the new theories stand up to this competition. In all these scenarios for the origin of life, there are a firm number of rules that the scenario must adhere to. The distinguished biologist Harold Morowitz has outlined these rules in the form of the following four "commandments":

1. Thou shalt not violate the laws of physics and chemistry, for these are expressions of divine immanence.

2. Thou shalt not trespass the Razor of Occam and multiply hypotheses, but shalt formulate the simplest of stories.

3. Thou shalt adopt a principle of continuity so that each stage of the grand scenario connects with the preceding and succeeding stages.

4. Thou shalt eschew miracles, for as Spinoza taught, they contravene the lawfulness of the universe.

With these precepts as our guide, let's review the bidding.

Out of the Soup

In the 1930s, Alexander Oparin in the USSR and J.B.S. Haldane in England independently conjectured that life on Earth got started when chemical compounds available 4 billion years ago fortuitously combined in the primeval seas to form objects that could metabolize or replicate. Thus, the central ingredient in the Oparin-Haldane scenarios was a physical environment consisting of what Haldane termed "a warm, dilute soup," in which the right kind of chemical reactions could take place to generate the "right stuff" for life. So all theories of this type have acquired the collective label "soup theories." All soup theories were given a boost in 1952, when Stanley Miller demonstrated that some amino acids—the essential components of proteins—could be generated from inorganic matter in an atmospheric environment similar to that thought to exist on the early Earth. Subsequent experiments have confirmed Miller's results, and extended them in various directions. So there is at least an experimental basis for believing in the feasibility of the creation of life by natural physicochemical means here on Earth. How *likely* is it? Well, that's another story; in fact, it is our story in this chapter.

Let's first summarize the "metabolism first" scenarios.

• *Oparin's scenario:* Basically, the Oparin view of life is that it began with what are called *coacervate* droplets, combinations of histone (a protein) and gum arabic (a carbohydrate). Adding enzymes (other proteins) and sugars to this mixture, Oparin came up with droplets that contained starches and that gr v larger until they finally split into daughter droplets (replication). So the basic structure of this theory is

Cells (Coacervates) → Enzymes (Proteins) → Genes

• *Fox's scenario:* Sidney Fox, a chemist at the University of Miami, advanced the notion that proteinoid microspheres are the answer to the origin-of-life question. The essence of this idea is that if you heat up dry amino acids in the presence of certain other amino acids, what you get is chains of amino acids (proteins)—but not the type that correspond to proteins occurring in an earthly biology. Nevertheless, Fox's proteinoids display certain features of living cells when dissolved in water, including the ability to grow (metabolize) and form many copies of themselves (replicate). But they do not possess any sort of hereditary mechanism, nor do they evolve by natural selection. In summary, then, here is Fox's scenario:

Amino acids → Proteinoid Chains → Cells → Genes

• *Shapiro-Dyson scenario:* Chemist Robert Shapiro of New York University and physicist Freeman Dyson of the Institute for Advanced Study in Princeton have independently put forward scenarios that have life on Earth beginning with some sort of primitive proteins that could metabolize, with a genetic information storage mechanism coming along *much* later. The basic thrust of these double-origin theories is that the genetic apparatus did not arise out of the original proteins, as with the scenarios of Oparin and Fox, but had an entirely different functional activity initially. But when it turned out that replication was a kind of side aspect of its original function, this replicative function proved to be so much better than the original protein replication machinery that it just took over. Diagrammatically, the Shapiro-Dyson scheme looks like this:

Cells → Proteins → → RNA → DNA

much later

• *The Cairns-Smith scenario:* The Scottish chemist Graham Cairns-Smith has proposed a novel origin-of-life theory asserting that in the beginning was simply . . . clay. The crux of this argument for how life got going is that crystals have the capability to make copies of themselves, as well as to grow by accumulating new material. And there's lots of clay around. So Cairns-Smith constructs a scenario by which the first life-forms are low-tech metabolizers *and* replicators constructed of clay. After these processes get going in silicon, carbon-based units take over owing to their much higher efficiency and the kind of life we see today gets started. Schematically, the scenario looks like this:

Clay → Growth/Replication → Organics → RNA/Proteins

takeover

Now it's time to turn to the genes-first scenarios, which postulate that the replicators came first, with the metabolizers the Johnny-come-latelies on the scene.

• *The Eigen scenario:* Manfred Eigen is a Nobel Prize–winning chemist from the University of Göttingen in Germany. He posits a sequence of steps in which at least one replicating RNA molecule forms by chance in the primordial soup. In some way, RNA molecules then learn to exert control over proteins, and a primitive genetic code develops. Now a cooperative set of interactions called *hypercycles* take place between the nucleic acids of the genes and the proteins, whose products take up the entire carrying capacity of the environment. At this point, competition enters again in the form of natural selection, with the hypercycles of chemical compounds each being protected inside its own membrane. Biological evolution then takes over from the chemical evolution of the hypercycles, and modern life-forms arise in an evolutionary fashion. Here is a diagram of the Eigen scenario:

Replicators → Hypercycles of Quasi-Species → Proteins → Cellular Membranes

• *The Gilbert scenario:* Harvard chemist Walter Gilbert posits an origins theory based on the observation that the RNA molecule is capable not only of storing information, but also of acting as an enzyme (i.e., a protein) under certain circumstances. This leads to a scenario in which the RNA molecules perform self-catalytic activity needed to assemble themselves from the "soup." These molecules then evolve in self-replicating patterns and develop a range of enzymatic activities, as well. In particular, the RNA molecules synthesize proteins, which push the self-catalytic activities of the RNA off center stage. Finally, DNA appears, giving a stable, error-correcting basis for information storage. So in this way, RNA is replaced by its creations, DNA and proteins, which now individually perform the functions that RNA alone initially carried out. A diagram of the process looks like this:

Self-Catalytic RNA → Proteins → DNA

Such are, in telegraphic summary, the competing Earth-based theories for how life got going, as things stood in the late 1980s. A discussion in far greater detail of each of these scenarios can be found in Chapter 2 of *Paradigms Lost*. Let's now turn to the off-Earth—but still natural—theories of the inception of life on the planet.

Not Invented Here

A vocal minority of investigators—including Nobel Prize winner Francis Crick—have put forth views stating that life as we know it did not commence here on Earth at all, but was imported from outer space. More specifically, the essential raw ingredients—microorganisms or spores from life-bearing planets elsewhere in the galaxy—are somehow transported across interstellar space to jump-start life on Earth.

The first proponent of this *panspermia* theory of life was the Swedish chemist Svante Arrhenius, who proposed it early in this century. His view was that the life-bearing spores sort of floated across space propelled by solar radiation. Later, Francis Crick improved upon this theory by suggesting that the spores were protected inside the shell of meteorites striking the primordial seas of the early Earth. In either case, this off-Earth origin theory solves the problem of how life got going simply by transporting the problem to another part of the galaxy.

An alternate off-Earth theory was put forth by Sir Fred Hoyle and Chandra Wickramasinghe, who noted that many important organic chemical compounds needed for life abound in vast clouds wandering between the stars. So they proposed a scenario by which cometary material "seeded" the primeval soup with all the right ingredients to bring life into existence.

And that's about it for the natural off-Earth theories. Now we come to the supernatural theory.

A Word from the Supernatural

Certainly one of the first "theories" of the origin of life is contained in the book of Genesis in the Bible. Strangely enough, several people like Duane Gish and Henry Morris, men with scientific training, have somehow managed to ignore the principles of that training long enough to advocate the creation sequence given in Genesis as being a literally factual account of how life on Earth originated.

In a book like this devoted to studies of *scientific* theories of observed phenomena, there is really no place for such "creationist" accounts of anything. For all the reasons put forth in *Paradigms Lost,* there is nothing even one bit scientific about appeals to supernatural forces like God as an explanation for why we see what we do and do not see something else. Nevertheless, we will touch briefly on developments over the last decade in the creationists' ongoing struggle with the forces of science, not as a serious contender in the origin-of-life sweepstakes, but as a wonderful example of how science and religion don't mix.

All of these positions—on-Earth, off-Earth, and supernatural—which have been explained in detail in *Paradigms Lost,* are summarized here in Tables 2.1 and 2.2.

LIFE ORIGINATED ON EARTH!

PROMOTER	ARGUMENT
Eigen	random replicators, hypercycles
Gilbert	self-catalytic RNA
Oparin	coacervates
Fox	proteinoids
Dyson, Shapiro	double origin, parasites
Cairns-Smith	clay

Table 2.1. Summary arguments for the Prosecution

LIFE ORIGINATED ELSEWHERE!

PROMOTER	ARGUMENT
"Natural Origins"	
Crick	extraterrestrial seeding
Hoyle, Wickramasinghe I	interstellar clouds and comets
"Supernatural Origins"	
Hoyle, Wickramasinghe II	silicon-chip Creator and diseases
Morris, Gish	creationism

Table 2.2. Summary arguments for the Defense

Before moving to consideration of what's new in any or all of these categories of origins theories, it's worth spending a few pages reexamining three of the most important aspects of living cells as we see them today: the genetic code, the so-called "junk" DNA, and the "handedness" of the DNA/RNA and the proteins (technically, their chirality), which tells us whether these molecules twist to the right or to the left in their three-dimensional geometry.

Three Hurdles on the Way to a Theory

The genetic code, the fact that 97 percent of DNA does not code for proteins, and the right-handedness of all nucleic acids used to form DNA/RNA and the left-handedness of all amino acids used to form proteins are all properties of the "high-tech" cell of today. These properties of modern life serve as touchstones for a viable origin-of-life theory, since such a theory should provide a means by which the primitive, "low-tech" form of a few billion years ago could develop into the structures we see around us today. So let's look at each property in a bit more detail.

The Genetic Code—and Its Pretenders

How does an origins theory account for the genetic code? The code is a dictionary that matches up a three-letter sequence of nucleic acids (codons) written on the cellular DNA with one of the twenty amino acids used by all
organisms alive today. This code is depicted in Figure 2.1, where each of the
sixty-four possible three-letter sequences of the four nucleic acids—
(A)denine, (T)hymine, (G)uanine, and (C)ytosine—is matched with the specific amino acid to which it corresponds. Note that three of the codons are
reserved for use as punctuation symbols, essentially telling the protein-manu-
facturing machinery to *Stop,* because the protein is finished.

Of special interest is whether this linkup between the alphabet of the
genes (the DNA) and the alphabet of the proteins (the amino acids) is accidental. Or did it arise in a special way so as to make modern life a lot more
likely to succeed?

There are about 10^{20} possible ways to match codons to amino acids.
But the one that nature actually uses (Figure 2.1) was put in place more than
3.5 billion years ago, very shortly after the actual origin of the first living things.
Recently, Steven Freeland, a graduate student at Cambridge University,

	T	C	A	G	
	phenylalanine	serine	tyrosine	cysteine	T
T	phenylalanine	serine	tyrosine	cysteine	C
	leucine	serine	*punctuation*	*punctuation*	A
	leucine	serine	*punctuation*	tryptophan	G
	leucine	proline	histidine	arginine	T
C	leucine	proline	histidine	arginine	C
	leucine	proline	glutamine	arginine	A
	leucine	proline	glutamine	arginine	G
	isoleucine	threonine	asparagine	serine	T
A	isoleucine	threonine	asparagine	serine	C
	isoleucine	threonine	lysine	arginine	A
	methionine	threonine	lysine	arginine	G
	valine	alanine	aspartic acid	glycine	T
G	valine	alanine	aspartic acid	glycine	C
	valine	alanine	glutamic acid	glycine	A
	valine	alanine	glutamic acid	glycine	G

Figure 2.1. The genetic code

and evolutionary biologist Laurence Hurst at the University of Bath have shown that it is among the best of this billion billion possible codes. They claim that natural selection acted very strongly on the space of possible codes, quickly weeding out those that didn't possess good error-correction properties.

The first thing that Hurst did was generate a lot of random codes and examine them for their ability to minimize the error caused by genetic mutations. He found that single-letter changes to a codon in nature's code, meaning that the wrong amino acid was inserted into a protein, tended to specify amino acids that were very similar chemically to the correct one, thereby minimizing the impact on the functioning of the protein.

Support for the idea that codon allocation evolved to minimize errors when RNA is translated into amino acids comes from the general observation that errors in base sequences within codons occur more commonly at those positions having least impact on the characteristics of a specified amino acid. For instance, the second codon position corresponds closely to the amino acid's affinity for water. Most "water-loving" amino acids have A in the second position, while most "water-hating" ones have U. This position is less prone to error than the third codon position, where mispairing has less drastic effects on the properties of the resulting amino acid.

Freeland's work followed almost immediately, accounting for errors occurring during the decoding of the gene in the process of translation. His reasoning was that if the code had evolved to minimize translation errors, it should minimize chemical differences most between the correct and incorrect amino acids at the third slot in the three-letter codon sequence. This is because the translation machinery misreads this letter ten times as often as the second letter.

Freeland's experiment showed that no more than one random code in a million was better at reducing the impact of errors than the natural code. A graphical depiction of these results is shown in Figure 2.2, where all the codes examined by Freeland are plotted against their susceptibility to error. The code that nature has evolved to use is very far to the left of the main cluster of randomly generated codes. It is very unlikely that such an efficient code arose by chance. Rather, natural selection must have played a very big role.

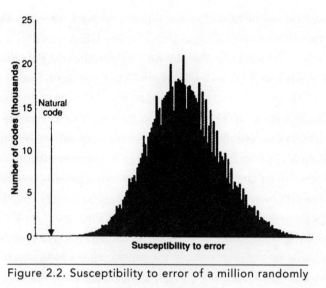

Figure 2.2. Susceptibility to error of a million randomly
generated genetic codes

The argument for the strong selection pressure is that there are good reasons why the code should be universal. That's because most organisms exchange DNA. For example, a virus that infects two species can transfer DNA from one to the other.

Talking Trash: The Strange Puzzle of "Junk" DNA

Reading popular accounts of the Human Genome Project sometimes gives the impression that by the year 2005, when every one of the estimated 100,000 human genes will be known, we will have everything in hand to enable us to understand the workings of the cell. What's wrong with this neat little picture is that the protein-coding portion of the genes account for only about 3 percent of the DNA in the human genome; the remaining 97 percent encodes no proteins. Most of this silent genetic majority has long been thought to have no function at all; hence its name: "junk" DNA. But there are now a growing number of scientists who believe that there are riches hidden in the junk DNA that will help us to better understand normal genome repair and regulation, as well as dis-eases—including cancer.

The stretches of DNA that do not code for proteins are called introns. You might think of them as a bit like the "commercial breaks" between the sections of the gene called the exons, where the protein instructions are found. Some have proposed that the introns are relics of genes that have become useless or redundant. Others say that introns encourage genetic diversification by allowing several proteins to be manufactured from one gene, using different combinations of introns and exons. But neither of these expla-nations has ever been substantiated.

The prevailing view at the moment is that the introns are there to make sure that the genetic information in the exons is processed properly, so that proteins made by the gene contain no imperfections. In information-theory terms, the introns are the error-correcting bits of the code. Another intriguing aspect of junk DNA is that it contains the hint of an actual language.

One type of empirical relation that seems to turn up frequently in com-plex systems studies is what's termed a *power law*. A particularly intriguing illustration of this kind of "law" is the relation observed between the rank order of words in a language and the frequency of the appearance of these words in a sufficiently large body of text. This relationship, now termed Zipf's Law, was first presented by George Zipf in his 1949 volume *Human Behavior and the Principle of Least Effort*. To describe this "law," suppose we list the words of the English language according to how common they are. So, for example, the most common word is "the," which we assign rank 1. The next most common word is "of," which is then given rank 2, followed by "and" having rank 3, "to" with rank 4, and so on. Note that word rank is simply an ordinal number, an integer representing the order of the most common words. Frequency is like a probability, measuring the likelihood of a particular word's occurrence in a large body of text. What Zipf discovered was that if we plot word frequency in a large body of English text versus the word rank, we obtain a graph of the type shown in Figure 2.3.

Algebraically, Zipf represented this relationship between word fre-quency and word rank by the expression

$$f(r) \sim \frac{1}{r \log 1.78\,R},$$

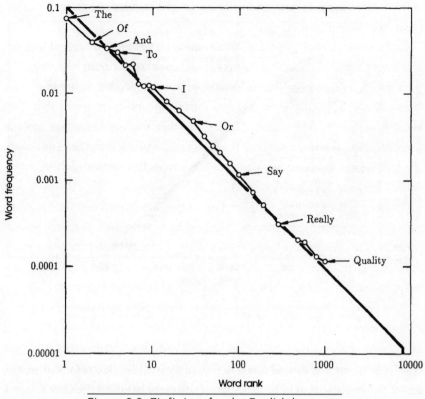

Figure 2.3. Zipf's Law for the English language

where *r* is the word rank and *R* is the number of words in the language. Relations of the form $f(r) \sim r^\alpha$ are called power laws, in this case an *inverse power law,* since here $\alpha = -1$. For a good writer with an active vocabulary of, say, $R = 100,000$ words, the ten highest-ranking words occupy 24 percent of a text, while for a popular novel or newspaper using a slimmed-down vocabulary of $R = 10,000$ words, this percentage increases marginally to nearly 30 percent. To illustrate an in-between case, Mr. William Bowers of Thousand Oaks, California, kindly sent me a Zipfian analysis of Conan Doyle's classic Sherlock Holmes story "The Hound of the Baskervilles." Using the same vocabulary size of $R = 10,000$ words, the result of Bowers's analysis of this 59,498-word story is shown in Figure 2.4. The striking conformance of the Conan Doyle story with Zipf's theoretical prediction is clear, although it's interesting to note that in this story the word "to," which holds rank 4 in gen-

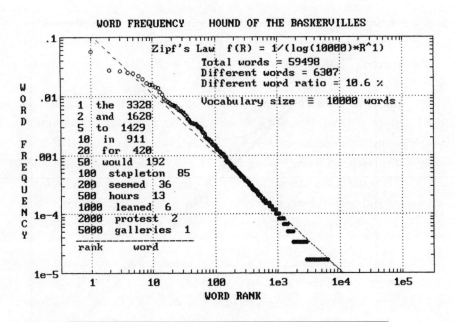

Figure 2.4. Zipf's Law and "The Hound of the Baskervilles"

eral English text, has moved down to fifth place in the rankings for this particular text. Note also that in his analysis, Mr. Bowers has used the scale factor 1 rather than Zipf's original value of 1.78, a change of no consequence insofar as the inverse-power-law relationship is concerned.

Zipf tried to derive the power-law form of his law by appealing to a principle of least effort. And, in fact, it's possible to show, using information-theory arguments, that by some process of selection, natural languages that survived are exactly those that were able to convey the maximum amount of information at a given cost, measured by the average time needed to produce the words of the language. Only those languages obeying the inverse power law above have this property. But wait! It turns out that monkeys hitting typewriter keys at random will also produce a "language" obeying Zipf's Law.

Suppose we consider a nine-letter alphabet with a space character, so that our mythical monkey strikes each character with likelihood 0.1. Many years ago, Benoit Mandelbrot showed that the exponent in Zipf's Law is $\alpha = -1.048$ for this monkey language, hardly any different from the value $\alpha = -1$ for English. But the median word rank in this language is 1,895,761, which means it takes this many of the most frequent words in the monkey language to reach a total probability of 0.5. By way of contrast, the analogous figure in English is between 100 words (for typical media texts) to 500 (for belles lettres writers). Thus, the monkey language is a very wordy one compared with English.

Somewhere between the language of English and that of the monkey is the language of DNA. Recently, physicist Eugene Stanley, in collaboration with a group of geneticists, applied the Zipf test to the part of yeast DNA that does not correspond to the coding region for any genes, the so-called junk DNA. When the researchers arbitrarily divided up the junk into "words" between 3 and 8 bases long, what emerged was a surprisingly tight fit to the theoretical Zipf curve with exponent $\alpha = -1$, as shown in the diagram in Figure 2.5. Moreover, the researchers applied a second test to quantify the "redundancy" in the yeast language, finding a level significantly higher than would be expected if the "junk" were completely random. These two findings together suggest that something is written in these mysterious regions, and that the junk may not be so much junk, after all. As Harvard biologist Walter Gilbert described it, "I think the junk is like the stuff in a junk shop. You can find lovely things in it." Finally, let's turn our attention to the strange puzzle of the "handedness" of the molecules of life—the nucleic and amino acids.

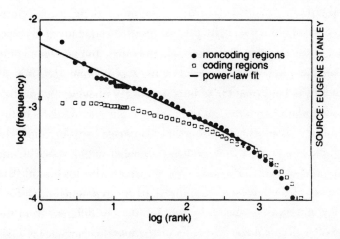

Figure 2.5. Zipf plot for arbitrary words in noncoding part of yeast DNA

Chirality: The Twists of Life

How did life on Earth come to use almost exclusively "left-handed" amino acids? In laboratory experiments of the Miller type, in which one generates various amino acids by simulating the primordial conditions, almost equal amounts of left- and right-handed amino acids are produced. Yet in the real world, life on Earth uses only the left-handed variety. It seems that life would not have arisen if nature had tried to work with both forms of its chemical building blocks, since experiments show that proteins cannot be built if both left- and right-handed DNA are present. So nature had to choose, one or the other, just as she had to choose one form of amino acid and let the other one go.

The best guess today at the answer to this puzzling query about the chirality, or handedness, of nucleic and amino acids is that some process favoring left-handed amino acids seems to have operated *before* the origin of life on Earth, and probably before the birth of the solar system. Thus, the answer, it seems, is literally in the stars.

One theory for how extraterrestrial events create left-handed amino acids was put forth in 1997 by Japanese researcher Yoshihisa Inoue and his colleagues. They mimicked the radiation from one hemisphere of a neutron star using a device that produced polarized ultraviolet light. The researchers then directed this radiation at a reaction in which the double bond joining two carbon atoms in a hydrocarbon flips around, giving a "twist" to the end product. So the product is left- or right-handed, depending on which way the double bond flips.

In normal ultraviolet light, the two products form in equal amounts. But when the Inoue group used their chiral ultraviolet, they produced more of one form than the other. The excess was only 0.12 percent, but they claim that their finding is confirmation of an extraterrestrial origin for the imbalance seen in life today.

Another Japanese group, headed by Kenso Soai at Tokyo University, offers an explanation for how even such a small initial imbalance can spiral into a virtual takeover. This team took a mixture of compounds containing a small excess of the left-handed amino acid leucine. In the presence of this imbalance, the components of the solution reacted to form a compound that also had a small imbalance of left-handed amino acids. But the molecules of this new compound acted as a catalyst in its own formation, and soon almost all the compound was left-handed. So the tentative conclusion is that polarized light in outer space created a small bias toward left-handed amino acids, and that chemical reactions later amplified this excess so as to essentially drive out all the right-handed ones.

With all these various preliminaries in hand about past theories, as well as some understanding of important sideshows pertaining to the genetic code, junk DNA, and chirality, we can finally move to a consideration of some of the exciting new ideas in the origin-of-life business. So without further ado, let's turn to perhaps the most exciting one of all, what's often called the *RNA world*.

In the Beginning Was ... RNA

In 1982, Thomas Cech of the University of Colorado shocked the world of biochemists by finding RNA (ribonucleic acid) molecules with catalytic powers in a single-celled protozoa called *tetrahymena*. A year later, Sydney Altman and his colleagues at Yale University discovered an RNA enzyme, now called a *ribozyme,* sitting inside the bacterium *E. coli.* This discovery was made by tracing the enzyme from its scissorlike ability to cut other strands of RNA.

Many researchers greeted these molecules as relics from an era when all enzymes (proteins) were made from RNA. This notion gave birth to the label *RNA world*—a period on Earth when life was dominated by RNA; that is, when RNA played the dual role of both gene and enzyme.

But not everyone is happy with the concept of RNA world. Leslie Orgel, who studies models of early evolution at the Salk Institute for Biological Studies, says that the importance of RNA has been exaggerated. The big problem, he claims, is that there has never been any direct evidence to show that RNA-based life is feasible.

Now researchers led by Jennifer Doudna at Massachusetts General Hospital want to change that. Their goal is to make an RNA molecule capable of two things: replicating itself and making copies of other RNA molecules that are not equipped for self-replication. They want to show that a variety of RNA molecules with different chemical properties can, in principle, create molecular offspring with the help of just one special, self-replicating gene.

The importance of demonstrating this is enormous, since if such cooperating molecules existed early in evolution, their enclosure within membranes may have led to simple, RNA-based cells, and ultimately to an RNA world.

Thus far, the Doudna group has succeeded in constructing an RNA molecule that can replicate a fragment of itself. Interestingly, their approach exploits self-splicing introns, just exactly the "junk" part of the RNA—and the very entities that Cech discovered in 1982 (and for which he and Altman received the Nobel Prize in 1989).

Additional evidence in favor of the RNA world was discovered by N. Nitta in 1998. This work showed that fragments of the ribosome (the object that actually strings amino acids together to make proteins) containing only RNA are sufficient to catalyze the peptide bond synthesis linking amino acids. Such a result suggests the possibility that early life systems could have emerged from a world in which RNA molecules coupled amino acids together to make primitive proteins.

But if one believes in the by-now-conventional wisdom of the RNA world, where did the RNA come from? RNA is a very complicated molecule, one that's hard enough to synthesize in a test tube under controlled laboratory conditions. So how could it have formed spontaneously some 4 billion years ago, when there was nothing on Earth but a kind of random, prebiotic chemical soup? This puzzle is so imposing that most origin-of-life researchers believe that some crucial concept is missing. What makes this impasse especially frustrating, says Gerald Joyce of the Scripps Clinic in La Jolla, California, is that the RNA world scenario explains so many things so well. Without it, one is left with the baffling chicken-and-egg paradox: Which came first, proteins or DNA? The RNA world says that neither came first—because both DNA and proteins are descendants of RNA.

One possibility for this RNA precursor was suggested by C. Böhler, P. Nielsen, and Orgel in 1995. In this view, a simpler self-replicating genetic system was constructed from flexible, acyclic, achiral nucleotide analogs. One chemical entity that fits this bill is peptide nucleic acid (PNA). While the details are too cumbersome and technical to go into here, Böhler et al. reported results showing that PNA can act so as to transcribe its detailed genetic information directly to RNA, and thus initiate the RNA world.

Probably the most significant implication of this work is that it demonstrates a link between information-bearing polymers of contemporary organ-

isms and those previously considered to be inappropriate for prebiotic studies. This greatly expands the repertoire for reenacting life's beginnings.

Another approach for the formation of the vast amounts of RNA needed
to underwrite the RNA world is suggested by the German physicist Hendrik
Tiedemann of the University of Hamburg. He notes that the nucleotide bases
and sugars needed could be produced from hydrogen cyanide and formalde-
hyde, for which plausible prebiotic synthesis mechanisms in the early Earth
atmosphere exist. But a ready source of free energy would be needed to drive
the cycles of replication and evolutionary selection of RNA chains until self-
catalytic ribozymes arose that could couple the self-replicating RNA system to
other, not so easily available sources of free energy like solar radiation.

Some argue that this free energy might come from volcanoes, lightning,
or even ultraviolet radiation. But Tiedemann's idea is different. He says that
the most likely source for the needed free energy is the continuing bombard-
ment of the Earth by meteorites. According to his calculations, large impacts
could have repeatedly vaporized not only the entire ocean of the early Earth,
but also enough rock to create the energy needed to get the RNA world going.
The details of this process are again a bit too technical for a book of this sort.
But the numbers seem to bear out Tiedemann's thesis.

So there we have it. A scenario—RNA world—that explains almost
everything about the origin of life except how the scenario could ever get
started! Since RNA world is so appealing for so many reasons, researchers
continue their quest for the "missing link" by which the prebiotic environment could give rise to some replicator that could then be subjected to a "hostile takeover" by RNA. As this smoking gun remains elusive, let's turn now to
a reexamination of some of the otherworldly theories of the origin of life, theories that claim life was imported here from outer space.

Life from the Stars

In 1871, the British physicist William Thomson (later Lord Kelvin) suggested
that the barren early Earth could have been first colonized with life by alien
organisms drifting through space inside rocks. Thomson's theory included
the notion that the rocks—and their passengers—could have come from a
planet that had been shattered by a gigantic impact with another celestial body.

A few decades later, the Swedish physicist Svante Arrhenius proposed an analogous idea, conjecturing that microorganisms could ride the solar winds from star to star. And so began what is the oldest class of theories for how life began here on Earth, theories claiming it *didn't* begin here on Earth but was imported here by meteorites from other parts of the galaxy.

It's important to recall that sources of organic molecules on the early Earth divide into three categories: delivery by extraterrestrial objects; organic synthesis driven by impact shocks from meteorites; and organic synthesis by other energy sources, such as ultraviolet light or lightning. Chris Chyba and Carl Sagan studied these possible energy sources for the early terrestrial atmosphere, and discovered that the energy from meteorite impacts either produced or delivered quantities of organic material comparable with those produced by other energy sources.

One of the more dramatic events relating to this impact theory was the recent announcement of the discovery of life on Mars. Of course, dramatic claims require dramatic evidence to back them up. So let's take a little longer look at this controversial announcement.

The Martian Meteorite

In 1984, geologists investigating the Allan Hills region of the Antarctic ice sheet picked up a four-pound meteorite named ALH84001 that later analysis showed originated on Mars. This fact in itself turned upside down earlier theories about such interplanetary messengers, theories asserting that intact rocks could never leave the surfaces of their planets, since the planets were simply too massive, their gravity too strong, for any known natural process to lift rocks into space. But the real drama started on August 7, 1996, when investigators claimed that they had discovered evidence for life on Mars from analysis of the contents of the meteorite.

What researchers led by David McKay of NASA's Johnson Space Center in Houston concluded is that squiggly little objects in the meteorite are microscopic fossil remnants of organisms that once had lived on Mars. But the loyal opposition pooh-poohs this idea, saying that there's nothing lifelike about the Martian "bugs." Rather, they are simply a trick of the eye abetted by the peculiarities of the powerful microscopes used to look at them. Researchers such as

John Bradley of MVA, Inc., in Norcross, Georgia; Ralph Harvey of Case Western Reserve University in Cleveland; and Harry McSween of the University of Tennessee argue that most of the putative microfossils are nothing more than narrow ledges of mineral protruding from the underlying rock, which under certain viewing conditions can masquerade as fossil bacteria.

The originators of the nanofossil idea say that you have to be very careful, and that they know these mineral structures are there. But that's not what they are calling the Martian bugs. So how can this debate be resolved? Bradley's results show that there are definitely nonbiological processes that can produce these "buglike" structures. But he can't prove whether the particular structures imaged by the McKay group are microfossils or artifacts. The latest studies of ALH84001 suggest that organic molecules on the famous rock were picked up on Earth during the rock's long (16 million years) stay in Antarctica. In the final analysis, it may simply not be possible to prove they are microfossils from Mars.

But suppose they *are* microfossils? Then things get very interesting very fast, since a whole plethora of questions immediately arise. And among those of greatest immediate concern is, Why are microorganisms on Earth and on Mars so similar?

Cosmic Seeds

The eminent British astronomer Sir Fred Hoyle is famous (notorious?) for his advocacy of the idea that life came to Earth via bacteria and viruses carried on comets, a notion "cosponsored" by his colleague Chandra Wickramasinghe of the University of Wales in Cardiff. These ideas were recounted fully in Chapter 2 of *Paradigms Lost*. What was not discussed there is Hoyle and Wickramasinghe's views of ALH84001, in particular their explanation for why the putative microorganisms in the meteorite seem so similar to their earthly counterparts. According to Wickramasinghe, this is easily explained by their theory: "The reason that primitive life on Earth and Mars were similar is that both planets were seeded by similar organisms." End of story.

The seeding idea would also explain the puzzle of how life got off to such a quick start on both Earth and Mars. Basically, because it arrived in an already nearly fully developed form, and only needed a hospitable environment to thrive. Just the kind of environment, presumably, offered by the early Earth.

But what about life on Mars today? Always the optimist, Wickrama-singhe says, "I feel confident that it still survives." As he points out, on Earth life survives under glaciers and in hydrothermal vents deep in the ocean. These facts constitute the lesson that life can survive extremes that no one guessed before. "So," he says, "if life got going in an environment that was even remotely like Earth, I can't see anything stopping it."

Others, like James Lovelock, originator of the Gaia Hypothesis, are not so confident. In Lovelock's models, life either flourishes after tailoring the environment to suit its needs or it peters out. If life had survived on Mars, it would have taken over the planet—it wouldn't simply be hanging on. But the one thing that almost all researchers agree on is that if life exists on Mars today, it will probably take a manned mission to the planet to find it.

Life on Mars is one thing. Seeding of the early Earth by viruses and/or bacteria from outer space is quite another. The panspermia idea of Arrhenius is based on the notion that microorganisms from elsewhere in the galaxy were transported to Earth by stellar radiation. A major problem with this idea is that stars emit enormous amounts of ultraviolet radiation, so that the life expectancy of bare organisms in a stellar neighborhood fifty times the size of the Earth's orbit is only a few days. But in 1996, J. Secker and his colleagues put forth the idea that perhaps these microorganisms could overcome this problem by shielding themselves inside small protective mantles of material, roughly half a micrometer thick. According to their calculations, this would be enough protection to enable small organisms like viruses to survive the rig-ors of interstellar space.

What could such a mantle be composed of? Well, ice is no good, since it has very poor opacity to the most damaging ultraviolet wavelengths. The next possibility is carbon in the form of a small grain of dust, having the composi-tion of a typical meteorite. This would be good protection against radiation—but the extra mass of such a dust layer creates a weight that is very heavy for stellar radiation to propel through space at a velocity great enough to escape the gravitational attraction of a star like our Sun at a distance equal to that of the Earth in order to get to us. So very intense stellar radiation would be needed to give the organisms a sufficiently large "kick" to send them off into interstellar space.

The most promising sources of such intense radiation are red giants, highly evolved stars hundreds of times the size of our Sun—and hundreds of times more luminous. They are bright enough to carry organisms into space and, being cool, they emit light in the red end of the spectrum, with little ultraviolet content. Moreover, they are very old, and thus more likely than younger stars to harbor life on their surrounding planets.

Secker and his group estimate the survival chances of radiation-driven organisms during interstellar transit, finding that the physical size of the viruses and bacteria is crucial. Smaller organisms can support a thicker dust shield while remaining light enough to be driven by radiation pressure, so they are more likely to survive.

But radiation damage is not the only danger that organisms would face on their long trip to the Earth. Violent phenomena like volcanic eruptions and meteorite impacts are the most likely means by which microorganisms could leave their home planet. But such events necessarily generate shock waves in the planet's atmosphere, destroying some of the organic material before it can leave the planet. Moreover, entry into the atmosphere of a new host planet poses another serious threat, since most of the traveling material would be entering the Earth's atmosphere at very high speeds.

Finally, in order to expose the organisms to their new environment, their dust jackets must be fractured by atmospheric drag during entry. But too much heat will destroy them. Luckily, there is a wide variety of speeds at which they are likely to enter. So some fraction of Earth-bound organisms will be slow enough to survive the plunge down through the atmosphere to the planetary surface.

Is such *radiopanspermia* really possible? Secker et al. calculate that a high proportion of radiation-driven, dust-clad organisms could travel twenty light-years in about a million years. Beyond that, interstellar radiation and cosmic-ray particles pose a serious threat to their viability. So to travel farther, microorganisms would need to establish life around a new star, which in time could then cast out new organisms. Life propagating in this way, then, would have had time to populate a significant fraction of our Milky Way galaxy.

To verify the radiopanspermia concept experimentally, we would have to detect live organisms in interstellar space. To do so by direct observation seems unlikely, although there are some encouraging indications in the detection of presolar extraterrestrial helium inside some meteor impact sites here on Earth.

In a related development the discovery of the amino acid glycine inside the galactic cloud Sagittarius B2 in 1994 shows that the basic components of earthly life can be manufactured in the dead of space. So perhaps it's not even necessary for microorganisms to have originated on the surface of a distant planet; possibly they can form inside such clouds and then be transported to Earth.

This raises the intriguing possibility that the primordial soup in which living things got their start was not the oceans of the early Earth at all, but the interior of vast galactic clouds. That would certainly go a long way toward explaining the apparent speed with which life appeared on Earth, barely a few hundred million years after the planetary crust solidified. It appeared so soon because already highly evolved organisms from those clouds seeded the Earth via the panspermia process.

On that possibly uplifting note, we leave the life-from-space theories and return to the most mundane of earthly-origin theories, the contention that, just as claimed in the Bible, life emerged out of garden-variety clay.

Modeling Clay—and Life

According to conventional wisdom, life originated in the primordial soup that contained all the ingredients necessary to form the long, information-carrying polymers (chains of nucleic acids) able to self-replicate, mutate, and evolve. By now, the taste of this soup seems pretty agreeable—at least with regard to a fairly broad experimental base. This is the essence of the RNA world theory outlined earlier. But there are critics who go for other courses on the menu as far as the actual *origin* of life is concerned.

One of the oldest arguments against the soup theory is based on the fact that hydrolysis in the primordial ocean will always limit the chain length of polymers, preventing them from forming long enough chains to set up a genetic system. To get around this difficulty, maverick theorists like the chemist Graham Cairns-Smith of the University of Glasgow have proposed that the polymer chains can be obtained if they form on a mineral surface instead of in a free, liquid solution.

The so-called "clay theory" of the origin of life was summarized earlier, and the details are found in *Paradigms Lost*. So there's no need to reelaborate the story now. The only important point to note is that this theory makes it far easier to see how life could have gotten started; according to it, there was no need for an unlikely and fortuitous juxtaposition of chemical and geological events, only simple chemical reactions involving readily available materials.

Several experimental results of different types tend to support this claim, suggesting that the polymers of life were more likely to have been baked like prebiotic crepes than cooked in a prebiotic soup. French crepes are prepared by pouring liquid dough over a hot stone plate, causing the dough to dehydrate and solidify. In the chemical analogy, dehydration corresponds to the condensation of the polymer chain, regardless of the required temperature. The physical base for this new recipe is similar to one conceived nearly fifty years ago by J. D. Bernal, in which minerals absorb and concentrate the organic building blocks of life.

So prebiotic chemistry can now enjoy all the advantages bestowed by the theory of the organic synthesis of long polymer chains. There is the additional advantage that a clay surface helps considerably in the survival of a polymer chain owing to the reduced rotational and translational movement within the chain. This makes condensation at surfaces more selective and perhaps somewhat slower than condensation in solution.

But even Cairns-Smith doesn't claim that these results bury the soup theories. The soup is still required ultimately to provide the building material. Thus, it is certainly compatible with a surface-mediated origin of life. The message is that the earliest forms of life may have proliferated by spreading on surfaces—and then migrating back into the soup. As Cairns-Smith says, "My position re the RNA world is that it was the final of perhaps a series of takeovers, which represented the exploratory mechanism for the evolution of the genetic system. So I'm all for it!—except of course that it should have been called 'the RNA suburb'!"

Some Liked It Hot

In 1996, researchers found a new form of life. A strange organism was discovered near a place on the ocean floor where superheated water pours out of a vent. This exotic life-form flourishes at near-boiling temperatures under pressures two hundred times greater than what we experience at sea level on the surface. Furthermore, all this beast needs to live is nitrogen, carbon dioxide, and hydrogen, from which it makes natural gas (methane).

By sequencing its genetic makeup, scientists have confirmed that this organism is in fact a member of a previously postulated form of life called *archaea.* Schoolbooks have previously recognized only two life-forms: prokaryotes (bacteria with no nucleus in their cells) and eukaryotes (everything from yeast to humans whose cells have a nucleus). Now there is a third type—the archaeon.

The genetic sequencing of an archaeon shows that some of its genes are closely related to bacterial genes, while others, involved in directing the cell's information processing, are similar to eukaryote genes. So it seems that the archaeon has a foot in both previously known domains and may well be their common ancestor. Craig Venter, leader of the team that sequenced the genetic makeup of the archaea, says, "I would be surprised if we don't find organisms similar to this at some stage on other planets." It's certainly plausible to imagine that archaea, or something similar first seeded the Earth's primeval hot oceans, carried here perhaps on the surface of a falling meteorite.

In this new kingdom are organisms that can live at freezing temperatures, while others thrive at temperatures beyond the boiling point of water. Still another can survive more than a million rads of radiation, making it the hands-down favorite to be the ultimate victor in a nuclear holocaust.

Of special interest is that the archaea kingdom does not consist of just a handful of oddball organisms. Rather, this previously unrecognized form of life constitutes an eye-popping one third to one half of all the biomass on Earth, most of it in the oceans. This fact, if no other, underscores just how little we really know about our planet. We know more about the surface of the Moon than about the deep oceans of the Earth. The idea that life-forms might have arisen near these thermal vents in the deep ocean forms the basis for another theory of the origin of life, what I'll call here the Iron Pyrite theory. Let's have a brief look at what it proposes.

Ironing Out the Origin of Life

The byline of a patent attorney is not what one usually sees in the journal *Systematic Applied Microbiology.* But, then, Günter Wächtershäuser is no ordinary patent attorney. He is also a trained chemist, who, like the gentlemen scientists of an earlier era, has kept his brain open and his mind active in research even as he wades through tedious tomes of patent applications. In 1988, Wächtershäuser proposed an answer to a big puzzle in the origin-of-life derby. In his paper, Wächtershäuser noted that certain bacteria that thrive in an oxygen-free environment have a strong chemically based metabolic activity. He went on to suggest a reexamination of the long-neglected view that the origin of life and the origin of such metabolic activity coincided. This contrasts with the conventional wisdom of the soup theorists, as well as with the view that the original metabolism was based on photosynthesis, not chemical activity in the sea.

Wächtershäuser suggests that the neglect of the chemically based metabolism option stems from the problem of finding a likely energy source to drive such a metabolism. His proposal is that this source was sulfur chemistry in an oxygen-free world. In this scheme of things, volcanoes produced iron and sulfur compounds, which reacted in water to form pyrite (fool's gold!) releasing energy in the process. He says that the reaction could have taken place on submerged rocks or in moist areas near a volcano. Wächtershäuser also notes that the reaction could help generate organic compounds from carbon dioxide, another important step in the evolution of life.

In 1994, some important experimental evidence was finally put forth supporting Wächtershäuser's case. A group at the University of Regensburg in Germany found that the pyrite reaction triggers the formation of bonds that are similar to the carbon-nitrogen bonds that link amino acids to form proteins. Although the scientists didn't form actual proteins, they did manage to demonstrate the sort of bonding needed for proteins to be produced. Additional experiments are under way to also generate the important organic compounds from carbon dioxide, the other big part of the Iron Pyrite theory. But at present it is still unclear how serious a contender iron is in the race to life.

Self-Replicating Long Shots

Proteins perform most of the chemical tricks that make life as we know it possible—with one important exception: They do not hold the blueprint for their own replication. That key piece of information is the province of the genes.

But recently chemist Reza Ghadiri of the Scripps Research Institute in La Jolla, California, built a protein that can replicate unaided. Now for the first time, hard evidence is available that life could have arisen solely from proteins.

As Ghadiri stated the matter, "As far as we know it, proteins are the best catalysts on Earth, and I saw no reason why they shouldn't be able to self-replicate." The only thing that put natural proteins at a disadvantage when compared with RNA was that they didn't act as templates for their own replications.

Ghadiri's solution to this problem was to construct a simple self-replicating protein. The procedure is to start with a natural protein from yeast called GCN4, in which two identical protein pieces, each thirty-two amino acids long, fit together like the two sides of a zipper. Ghadiri reasoned that one side might be able to act as a scaffold for the other, thus forming a template in its own reproduction from simpler components.

To test this idea, Ghadiri's team split GCN4 molecules into their identical halves, which they chopped into two chunks. They then mixed these chunks at room temperature to see if they would join up again. Because there was no full-size template around, the only way the fragments could do this was to find each other and link up by random motion. For that reason, the complete thirty-two-amino-acid molecule re-formed very slowly at first. But the reaction got faster as it went on. According to Ghadiri, the only way this can happen is if the product was helping its own synthesis. Once the first full-sized protein piece was created, it acted as a template to accelerate the creation of the next. This pair then fell apart and each piece acted as a template for the next generation, and so on.

While people gasp when they hear about this astonishing experiment, not everyone is convinced that it necessarily shows that life started with proteins. The problem is that while *every* piece of RNA can serve as a template, the type of protein that Ghadiri observes may be a rare phenomenon unlikely to arise outside a controlled laboratory environment. Ghadiri responds by saying, "I don't believe I'm that lucky. How could I have picked the one peptide in a million that can replicate?"

Other chemists have also been exploring molecular species possessing the all-important self-catalytic replication property. For example, Douglas Philp at the University of Birmingham in the UK has found four families of dimers (two molecules hooked together) that look promising as autocatalysts.

One of the most successful dimers, for instance, consists of an amine and an aldehyde, each of which has a site that "recognizes" the corresponding single molecule. Such a lock-and-key arrangement permits the molecules to stick to the dimer. If they stick in the right orientation, they can then react together to form a new dimer, and split off from the original one.

But the fact remains that a genetic molecule would almost certainly have to be a polymer (many molecules) if it were going to carry a significant amount of information. It would also have to be more chemically variable than anything Philp's team has come up with so far. Nevertheless, the work is intriguing. It suggests that RNA may not have had such an easy time of it at the beginning. Moreover, living systems based on biochemistries radically different from that which emerged on the Earth some 4 billion years ago are conceivable.

There is still one more scenario that merits attention before we move to the otherworldly contenders. This is a picture of the origin of life asserting that before any kind of living entity could have a chance of getting off the ground, there first had to be some kind of protomembrane in place to enclose the chemical reactions taking place. Harold Morowitz, a biologist from George Mason University in Fairfax, Virginia, is the principal proponent of this idea. Here is the basic outline of what he has in mind.

Membranes First

Every schoolchild knows that oil and water don't mix. But there are exceptions. For instance, soap removes grease from dirty clothes by floating it away in water. These exceptions always involve structures called *amphiphiles*, which are molecules having the property that one end is in oil while the other is in water. These amphiphiles were among the molecules brought to the surface of the early Earth by the action of solar ultraviolet light on molecules of earthly origin.

Amphiphiles were not completely at home in the primordial sea, since when these molecules collide the oil-seeking parts adhere to each other, while the water-seeking ends interact with the surrounding sea. This results in collections of amphiphiles called *coacervates*. Membranes are coacervates made of amphiphiles bonded together in sheets two molecules thick. The oil-seeking ends form the interior of the sheets and the water-seeking ends the exterior. In conformance to the basic laws of chemistry and physics, these sheets then spontaneously fold up into closed shells called *vesicles.*

So when the first vesicle occurred, the membrane became a barrier separating the water inside the vesicle from that on the outside. This is the beginning of individuality. And while all the chemical reactions were taking place in the primeval soup, occasionally a molecule was produced that could stick to the membrane surface and catalyze one or more reactions going on inside the vesicle. Those molecules that catalyzed more rapid chemical reactions led to faster growth of the favored vesicles in which they were found. So the earliest vesicles were thus characterized by a network of chemical reactions. This was the beginning of the metabolic basis of life.

So the overall scenario that emerges from the Membrane Theory is the following:

primordial chemistry → amphiphiles → vesicles →
amino acids → nucleic acids → genetic code

This completes our account of the scientific theories of the origin of life—earthly and otherwise. But there are still the mystical contenders. Normally in a book of this sort such "creationist" theories would be passed over in silence, in obeyance to Morowitz's Fourth Law enunciated in the opening section to the effect that one should eschew miracles. So the material of the next section is offered more for its role in the sociology of science than for its content as a viable theory for the origin of life. It is not a theory, scientific or otherwise; it is simply a scenario.

The Struggle for Souls

Paradise, California, seems to be a paradise for conservative Christians who want their children's science education to include an account of the origin of life based on the book of Genesis. At a new school there, a sympathetic board of directors announced recently that it plans to let the creationist parents have their way. Similar schools in Michigan and Nevada may follow suit.

In 1987, the U.S. Supreme Court struck down laws requiring equal time for creationist and mainline scientific accounts of evolution and the origin of life. Yet nine years later, the struggle over evolution in the schools is alive and well. While the legal setbacks of the 1980s left their mark on the antievolution movement, now instead of lobbying for state laws to put "creation science" into the classroom, advocates are returning to the grass roots in their campaigns. They are putting pressure on local school boards and teachers trying to make evolution a topic that's "too hot to handle."

What is it *exactly* that makes well-meaning people fight so fiercely to keep children from learning such a basic scientific principle as evolution? From the beginning of the creationist movement, the driving force has been the same: the struggle for souls. Students who learn evolution, say the creationists, will come to doubt the existence of God. And without that moral rudder of religion, they will become bad people doing bad things. Thus, evolution is evil and a cause of evil. Indicating the dimensions of this argument, Henry M. Morris, probably this century's most influential creationist, wrote that "evolution is at the foundation of communism, Fascism, Freudianism, social Darwinism, behaviorism, Kinseyism, materialism, atheism and, in the religious world, modernism and Neo-orthodoxy." Wow!

In *Paradigms Lost,* the story of the famed Arkansas case *McLean* v. *Arkansas Board of Education* was spelled out in some detail, especially since the decision striking down the state's antievolution law by Judge William R. Overton gave such a clear-cut account of the difference between science and religion. In particular, the judge wrote, "a scientific theory must be tentative and always subject to revision or abandonment in light of facts that are inconsistent with, or falsify, the theory. A theory that is by its own terms dogmatic, absolutist and never subject to revision is not a scientific theory." End of story—or so hoped the scientific community. But not so, said the creationists.

In 1987, U.S. Supreme Court Justice William Brennan wrote, in his opinion on the Court's review of the Louisiana case *Edwards* v. *Aguillard,* that "teaching a variety of scientific theories about the origins of humankind to schoolchildren might be validly done with the clear secular intent of enhancing the effectiveness of science instruction." Justice Antonin Scalia in a dissenting opinion on the same case opened the door for today's assault on science by saying, "The people of Louisiana, including those who are Christian fundamentalists, are quite entitled, as a secular matter, to have whatever scientific evidence there may be against evolution presented in their schools."

Sad to say, it appears that Justice Scalia failed to foresee that his phrase "scientific evidence . . . against evolution" would be one of the main euphemisms antievolutionists have devised to avoid referring to creation. Authors of "creation science" books comb scientific journals looking for anomalies, then construe them as suggesting that evolution never took place. An excellent example is Moon dust. According to an old *Scientific American* article, if the Moon were really billions of years old it should be covered with several hundred feet of dust from the assumed amount of meteorite dust falling on it. But the *Apollo II* astronauts found the dust to be only a few inches deep. Therefore, the solar system must be no more than a few thousand years old—too young for evolution to have taken place. Or so goes the argument. But satellite measurements have shown that this old article vastly overestimated the fall of cosmic dust. Still, Moon dust remains a weapon in the creationists' arsenal (since they are never willing to revise their theories).

One may well ask: What's wrong with alternatives to evolution? Why shouldn't students hear all sides of a controversy?

Well, what's wrong is that there is no controversy. Good critical-thinking exercises should deal with issues that are actually in contention. It certainly is true that evolutionary mechanisms, rates, and phylogenies are being debated in scientific circles; whether evolution happens is not. You might as well try to repeal Kepler's laws of planetary motion or the heliocentric model of the solar system. Forced to choose between religious faith and a badly understood scientific principle, how many of the general public will choose evolution? Not many, I'd say.

But this is a false dichotomy. Some of the strongest criticism of creation science has come from mainstream Christian denominations, who see evolution as part of God's grand plan. Stripped to its essentials, the difference between creation and evolution boils down to a question of history: Does the universe have a history? Or was everything in it created as it is, all at once?

So there it is. The same old creationist arguments leading to the same old conclusions. Most Americans have already made their choice to be religious. Now you must choose which you prefer—a religious population that accepts evolution or a religious population that rejects it.

THE APPEAL:
SUMMARY ARGUMENTS

There is a rich array of possibilities on offer from the Prosecution in the argument for reopening the case for life originating on Earth through natural physicochemical processes. RNA world, the sulfur chemistry of thermal vents, membranes, self-replicating proteins, and all the rest. But the new evidence favoring the Defense also has something to recommend it, especially the notion of radiopanspermia. Table 2.3 summarizes the main new evidence, pro and con.

EVIDENCE	INVESTIGATOR(S)	EVIDENCE FAVORS
RNA world	Joyce, Doudna, Tiedemann, Nitta	Prosecution
Iron sulfides	Wächtershäuser	Prosecution
Thermal vents	Venter	Prosecution
Clay minerals	Cairns-Smith	Prosecution
Membranes	Morowitz	Prosecution
Self-replicating proteins	Ghadiri, Philp	Prosecution
Radiopanspermia	Wickramasinghe, Secker	Defense
Creationism	Gish, Morris	Defense

Table 2.3. The evidence

THE DECISION: APPEAL UPHELD

A decade ago, I cast my vote for the Prosecution's clay theory of the origin of life. Even then this scenario was a bit of a long shot. But sometimes in life the long shots do romp home ahead of the pack. In the intervening decade, though, it doesn't seem as if much new evidence has appeared—for or against—Cairns-Smith and the clay theory. The same cannot be said of several of the competing Prosecution scenarios, especially the RNA world theory based on the self-catalytic properties of RNA. On the other side of the fence, the Defense has not really mustered much by way of compelling evidence in support of its off-Earth scenarios. So, on balance, this time around it looks to me as if the way to bet is with the favorite, the RNA world. While not without its own set of difficulties, these obstacles look far less imposing than the barriers in the path of any of the competition.

THREE

Genetic Imperialism

Claim: Human Behavior Patterns

Are Dictated Primarily

by the Genes

BACKGROUND

Nature-Nurture, Sense or Nonsense?

In the summer of 1993, Dean Hamer and his colleagues at the National Cancer Institute in Washington, D.C., lobbed a bombshell into the midst of those who feel that human social behavioral patterns are principally generated by their social environment. What Hamer announced was the result of a study of 40 pairs of brothers, both of whom declared themselves to be exclusively homosexual. The study showed that the DNA of 33 of the 40 pairs showed a significant correlation between homosexual behavior and the inheritance of a genetic marker near the end of the X chromosome.

So does this result mean that there is a gene for homosexuality hidden in that DNA? Previous studies conducted at both Tufts University and Northwestern University in 1991 appear to also contradict the widely held perception that sexual orientation is largely determined by a child's early influences, conditioning, and environment, and to suggest a biological foundation for homosexuality. As studies like these continue, the weight of evidence still favors the notion that homosexuality is more of a predisposition than it is a preference.

The homosexuality question strikes right to the heart of the sociobiology question: Are human social behavioral patterns determined all—or even mostly—by our genetic makeup? Or, alternately, does the environment and culture in which we are brought up shape the way we behave? To put it in everyday terms, does nature or nurture dominate the way we interact with our fellow human beings?

When sociobiologists first suggested in the 1970s that homosexuality might be biological, they were called Nazis and worse, reflecting the way that Nazism attempted to produce genetic explanations of human social activity. Economic growth following World War II encouraged most Western governments to imagine that they could eliminate social problems by mixing enlightened planning and generous spending. In short, they believed they could steer human nature.

In such an atmosphere, sociologists made lucrative careers espousing "nurture" explanations for everything from educational failures to religious inclinations. Rather than risk being accused of a fondness for jackboots and martial music, geneticists of the time stuck to studying fruit flies and honeybees. But the fashion began to change in the 1970s, the catalyst being Edward O. Wilson's Pulitzer Prize–winning volume *Sociobiology*, which was published in 1975.

In this work, Wilson argued that there is no reason to exempt humans from the types of genetically determined social behaviors seen in the animal world. And so things like altruism, homosexuality, religion, and the like were all of a piece, and could be studied using the same biologically based methods used in the worlds of animal ethology and molecular biology. The howls of outrage from the academic Left at the time Wilson put forth his arguments were heard from Cambridge, Massachusetts, to Cambridge, England. Wilson himself was accused of being everything from a charlatan to a Nazi, and was essentially banished from polite scientific company for his heresies against the received wisdom of the day. At that time, this ultraleftist political view of genetic determinism worked to cement the view that human behavior and genetics could not be linked in a scientific hypothesis to be explored. But times have changed, and recent work has shown that there are strong connections between the way we behave and the way our genetic pattern is structured.

Chapter 3 of *Paradigms Lost* provides a full account of the merits and demerits of the gene-based view of human nature and behavior. So let me just repeat the summary arguments for and against this position.

THE LOWER-COURT VERDICT

A decade ago the competing positions on the sociobiology issue looked like this:

HUMAN BEHAVIOR IS PRIMARILY GENETIC!

PROMOTER	ARGUMENT
Lorenz	innate aggression, group selection
Wilson, Barash	genetic influence, multiplier effect
Dawkins	selfish genes
Lumsden and Wilson	coevolutionary circuit
Trivers	reciprocal altruism
"Theoretical Support"	
Hamilton	inclusive fitness, kin selection
Maynard Smith	evolutionary game, ESS
Axelrod	evolution of cooperation and norms

Table 3.1. Summary arguments for the Prosecution

The foregoing table has been split into two pieces, one offering direct evidence in support of the thesis that the genes determine most of what's important about our behavior; the other listing a variety of theoretical ideas lending support to the empirical evidence. Most of this theoretical support comes from the field of game theory, which is devoted to understanding the best ways for *rational* agents—humans, animals, corporations—to resolve conflicts. Both the direct evidence just noted and the theoretical work will be explored in more detail in the balance of this chapter.

The other side of the genetic determinism coin argues that most important human behavioral patterns are culturally determined, not the product of the accidents of biology that created our particular DNA pattern. When it comes to arguments favoring the social climate as the determinant of human behavior, the supporting evidence a decade ago shaped up as follows:

HUMAN BEHAVIOR IS PRIMARILY ENVIRONMENTAL!

PROMOTER	ARGUMENT
Boston Group	reification, no multiplier effect, unfalsifiability
Schwartz	evolutionary constraints
Sahlins	kin selection impossible
Gould	*Just So Stories*
Dawkins	cultural memes

Table 3.2. Summary arguments for the Defense

With these positions on the record, let's now turn to an examination of what's happened in the sociobiology business in the past ten years.

THE APPEAL

It's in the Genes

Can variations in a single healthy gene be all it takes to create differences in how people behave? Well, consider the following tale of the prairie and montane voles, mouselike rodents inhabiting the American Midwest. The prairie voles are devoted to their mates and offspring, and rarely take up housekeeping with new partners—even when one dies. Couples often raise multiple litters together, and they seldom kick their offspring out of the burrow.

By way of contrast, the montane voles, who live in the Rocky Mountain regions, form few, if any, social bonds. They are promiscuous and heartless, living in solitary burrows, and abandon their young shortly after birth. How can one account for these diametrically opposed behavior patterns in the very same species?

Recently, Thomas Insel at Emory University in Atlanta discovered a single gene that encodes a hormone receptor in the vole's brain. According to Insel, this receptor governs the voles' behavior toward their families, as well as their overall pattern of social interaction. Similar studies on fruit flies showed that flies with one version of a particular gene tend to wander far from home, while those with another version tend to stay put.

Many geneticists, however, are skeptical in the extreme about the possibility of this "one gene, one behavior" view of social interactions. They point out that for many behaviors there will be dozens or hundreds of contributing genes, and all of them may exert small effects to the overall observed behavior. But they add that some behaviors will be caused by *major-effect genes,* a term coined by Dean Hamer, whose work was noted earlier. But discovering these major-effect genes may be as tricky as listening to a symphony and trying to hear the contribution of each instrument—with a lightning-and-thunder storm going on outside.

So why do people like Insel believe that their research puts them on the trail of a major-effect gene for a complex human behavior like social bonding? In the work on hormone reception in voles, Insel compared the brains of the female prairie voles and the female montane voles, finding that the receptors for the hormone oxytocin were located in completely different parts of the brain in the two species. This fact suggests that the pattern of oxytocin receptors in the brain plays an important role in female bonding with her offspring and her partner. As a check, Insel and his colleagues tested the hypothesis that the differences in behavior were related simply to the levels of the hormone itself. But they discovered that no matter how much oxytocin you give a montane vole, she won't become monogamous. "That's because her receptors are in an area of the brain devoted to scratching, not to pair bonding," says Insel.

After dismissing the level of oxytocin as a possible cause of the behavioral differences, the Insel team compared the oxytocin receptor gene from the prairie and montane voles, discovering slight differences in one part called the promoter. This serves as the dial that turns the gene's activity up or down, thus determining which cells have the receptors. Insel now conjectures that this is why the female prairie vole is faithful to her family and the female montane vole isn't. With different promoters for the oxytocin receptor genes, they are turned on in different parts of the brain in the two species. As a crosscheck, Insel carried out parallel tests on a gene for the receptor for vasopressin—a hormone that plays a role in the male vole's bonding with mates and offspring. Those tests showed similar results.

Now all this business about hormone receptors and social behavior in voles is fine. But what about humans? Do we also have genes for oxytocin and other hormone receptors in our brains that determine whether we're going to be faithful to our spouses or end up as loners? In the vole studies, Insel was looking at two species that evolved different versions of the genes and different behaviors over a period of tens of thousands of years. But humans are a single species made up of individuals with bonding behaviors ranging from lifetime monogamy to rampant polygamy.

These studies focus on social bonding. Let's look at another area of human behavior—our mental abilities—that also seem to be directly affected by our genetic makeup.

The Genetics of Brainpower

Suppose you want to measure the contribution of genes to general cognitive capability. So you propose an experiment in which human beings are cloned, in order to ensure that one of the groups you test is 100 percent genetically identical. It's not hard to imagine the howls of outrage from every corner of the ethical and political spectrum at this intrusion into nature's way of doing things. Yet, strangely enough, nature has been doing *exactly* this for centuries by creating identical and fraternal twins. And scientists from the time of Francis Galton have been using this natural "cloning" to study both diseases and traits such as blood pressure and intelligence quotient (IQ) that vary over a continuous spectrum of values rather than being all-or-nothing quantities. In

1997, a joint U.S.-UK-Swedish study based on identical twins reported the counterintuitive result that the genetic contribution to cognitive capability is remarkably constant through the course of a human lifetime. Since this study provides strong ammunition for the thesis that genetics forms the major contributing factor to intelligence, let's have a little longer look at what the researchers did and what they found.

The basis of the study was a collection of 240 pairs of Swedish octogenarian identical twins. The researchers determined the cognitive abilities of these twins by the classical method, which consisted of measuring the genetic contribution to intelligence at a single point in time. This is important, since an individual's performance on intelligence tests can be expected to vary considerably over the course of even a few years. So to make valid comparisons, it's necessary to standardize things at a given point in time. For these 240 pairs of twins, the proportion of trait variance attributable to genetics (the heritability) of general cognitive ability was 62 percent. This is a value remarkably consistent with the value of this same parameter seen in studies of adolescents and middle-aged adults. What is surprising and counterintuitive about this observation is that it had been previously supposed that with accumulated experience over time (nurture), the contribution of one's genetic makeup to intellectual functioning would decline. Now it seems as if it remains rather stable.

Complaints about the cross-sectional samples of twins at a single point in time have been addressed by studies measuring intelligence at various points in time. These tend to complement existing evidence for the strength of genetic influences on cognition. An overview of this data suggests that genes account for 0.50 of the variance in general cognitive ability, while shared environmental agents account for 0.33, nonshared environmental agents for 0.17, and measurement error for 0.10 of the variance in cognitive traits. Hence, the genetic agents are by far the most important component in the determination of intellect and its closely related traits, such as working memory and spatial skills. But this is not to say that environmental factors are negligible; they are just less important than the genes.

Homosexuality, social bonding, intelligence, are all part and parcel of human social activity. And the genes seem to be strongly implicated in generation of all these behaviors. But what about the other side of the coin? What evidence has been presented in support of nurture as the dominant force in generating behavior? Let's have a look.

Nurture Knows Best

Like all scientists, primatologists have a few bits of conventional wisdom that compactly summarize much of their field. These include the beliefs that a female macaque monkey's birth order is inversely related to her rank in the group, that orangutans are solitary, and that male chimpanzees remain with their birth group for life. New studies suggest that all of these old rules of thumb may have to be reexamined. The evidence coming from this work indicates that nonhuman primates exhibit a greater range of behavior than was previously thought, and that the environment—particularly the accessibility of food—plays a major role in determining behaviors that were previously thought to be largely innate.

An example of the type of work giving rise to the nurture-over-nature movement is that of David Hill of the University of Sussex in the UK. Hill has been observing a troop of Japanese macaque monkeys living under completely natural conditions on Yakushima, an island off the southern tip of Kyushu, the southernmost main island making up the country of Japan. Most of the previous studies of these monkeys involved troops that were *provisioned* (fed by humans) but otherwise lived in the wild. This feeding difference appears to have had a critical impact on behavior.

Researchers had noticed that provisioning brings animals in closer proximity to each other and results in more aggression. For instance, in the competition for food, a mother may be more protective of her youngest daughter, since that daughter is the most vulnerable one in this competition. Such special attention then translates into a higher ranking for younger siblings within the troop, and that ranking in turn follows those animals into maturity, not only in the area of feeding but others such as awareness of aggression from other animals. Past work has speculated on the possible evolutionary advantage conferred by this so-called *youngest ascendancy*.

Hill's Yakushima troop, however, showed a wide disparity of behavior by youngest daughters in their foraging for leaves and fruit. This implies that the mother's presence is not necessary to ensure that her youngest daughter gets her share. In Hill's words, "We didn't see any evidence of youngest ascendancy." One might summarize the principle at work here by saying that if you don't see any evidence of an effect, then perhaps it's a bit early to begin looking for the evolutionary argument underlying it. Other detractors note that one relatively short-term study like that by Hill is not enough to overturn decades of previous research. But it still raises the question of whether the youngest-ascendancy effect is simply due to the provisioning of a troop or is genetically based. It also serves to introduce the basic question of whether the kind of society we live in can influence the genes that are passed along to our children and grandchildren.

The Unselfish Gene

In the early 1960s, Oxford biologist Richard Dawkins popularized the so-called *selfish gene theory,* in which he claimed that evolution works not at the level of the group or even the individual, but at the level of the smallest element of life, the gene itself. This *genetic imperialism* was embraced by sociobiologists as providing a theoretical foundation for much of the work supporting the nature-over-nurture work of the time. In Dawkins's view, the organism is passive when confronting the environment, in the sense that the environment sets various challenges to the gene, which it then either passes or fails.

Steven Rose of the Open University in the UK argues otherwise, claiming that organisms select and transform environments by learning new practices and creating new cultures. Most term this *gene-culture coevolution,* which is a tricky mix of the interaction between the two strands of inheritance—genes and culture. The actual mechanism by which culture alters genetic makeup seems difficult to pinpoint. But there do appear to be instances where cultural traditions have acted upon genetic selection. For example, Marc Feldman of Stanford University offers the example of drinking milk. He says that humans can digest cow's milk only if they can produce the enzyme lactase, which allows absorption of the milk sugar, lactose. In those populations where cow's milk has been drunk for three hundred generations or so, up to 90 percent of the people have this enzyme. In groups that don't have a history of dairy farming, 80 percent of the population carry a different version of the enzyme and have great difficulty in drinking milk. Evidence of similar genetic changes stemming from cultural practices seems to be emerging in the animal kingdom, as well.

Zoologist Joseph Terkel of Tel Aviv University has been studying the changing behavior of Israel's black rats. Over the past century, much of their traditional forest habitat has been destroyed and replaced with plantations of Jerusalem pine trees. Terkel and his colleagues noticed groups of black rats living among the Jerusalem pines and surviving on seeds extracted from the cones. This would not be surprising but for the fact that cone stripping is a complex process, requiring a precise series of manipulations. So how did this cultural practice of cone stripping originate and how did it spread to an entire population in less than a century?

After more than a decade of study, cone stripping is still a mystery. Hungry adults don't spontaneously strip cones; neither can they learn the behavior by watching other rats do it. Only a small number of individuals are capable of the innovation needed to generate this practice but, once established, it spreads readily to all the rats in the population, passing from mothers to their pups. The key element seems to be the playful "theft" by the pups of partially opened cones from their mothers. These seem to give the young insight into the skill they must later use to work out how to begin the stripping process. So again, there appears to exist here a culturally determined behavioral trait (cone stripping) giving rise to a genetic drift (the theft of partially stripped cones by pups).

As a final example of the gene-cultural coevolution, consider the case of handedness. Why does it turn out that about 10 percent of Western populations are left-handed? For years, the standard explanation was that it is all in our genes. But recent studies have shown that growing up in a particular society at a particular time seems to have a powerful impact on the rates of left-handedness. A century ago, only about 2 percent of the population of North America were left-handed; now the rate is around 12 percent. Yet in Taiwan the rate remains at less than 1 percent. Work by Kevin Laland of Cambridge University leads to the conclusion that our genes bias the "handedness dice" to favor right-handedness. But conditions of early development—cultural effects—then fix the handedness at either right or left. Laland suggests these effects are things like mothers who actively encourage their infants to imitate them, or other mothers who actually punish their children if they use their left hand.

Memetics

One of the most prominent aspects of the nature-versus-nurture debate is the notion of a "cultural gene," or what Richard Dawkins has termed a *meme*. The basic idea is that just as material genes contain the blueprint for an organism's physical makeup and (possibly!) its subsequent behavioral repertoire, the information pattern inside brains serves as a kind of gene to produce certain types of culturally learned responses to different situations. A meme, then, is an element of culture passed on by imitation.

But Dawkins's memes differ substantially from the gene-culture coevolution scenario. They are not independent entities that shape evolution in the same selfish and mechanical way that genes do. The driving force behind a meme is not individual organisms or giant cultures, but little bits of information that are struggling to get replicated. For example, a typical meme might be a passage from a popular song or a religious ideology, both of which exist in different human minds and try to propagate themselves into other minds by co-opting the media or by word-of-mouth proselytizing. In this way, each meme is in competition with other memes for "resources," namely, neuronal space in the individual human mind. The winning memes then push aside the losers, thereby gaining an advantage in the race to see what aspects of a culture are propagated into future generations.

One particularly persistent type of meme involves religion. Celebration of the birth of Christ on December 25 is an especially good example. Christians hit on the idea of celebrating Christmas on this day at some time in the fourth century A.D. They really had no idea of the day on which Christ was born, but the date December 25 was already a pagan festival day, which marked the birth of the sun god at the winter solstice. At the time, the Christians were in a power struggle with the long-established Roman religion and needed something to keep their followers busy while everyone else was enjoying a holiday. So December 25 has been passed down from antiquity to the present as a feast of the nativity.

This example of the celebration of Christmas as a cultural rule, or symbol, that has been passed down through generations in societies culturally descended from Rome is exactly the type of thing that tempts us to extend the theory of evolution from the genetics of biological organisms to human culture. Even the way early Christians schemed to select the festival date reminds us of the struggle for existence that drives biological evolution.

Dawkins's notion of the meme, then, is simply an expression of the analogy between genetic and cultural evolution. Geneticist Mark Ridley points out that the idea of religious concepts as a kind of "virus" of the mind holds interest for another reason: It assumes that it's possible to distinguish between cultural parasites and everything else in a culture.

Consider a cultural change like the spread of television. It occurred at the expense of other modes of media communication—newspapers and radio, for example. But does that make TV a virus of the eyes and ears? Or is it just a normal cultural evolutionary change? From the standpoint of a newspaper owner, TV may seem like one giant cultural tapeworm. The television industry channels cultural energies, which would otherwise be consuming newspapers, into another medium, just as a tapeworm sucks energy out of its human host and converts it into more tapeworms. But we could also see the development of television as merely another cultural change of the kind that is regularly taking place in every culture. In this view, the emergence of TV is just like the famous case of the light-colored moths evolving into darker ones, in response to environmental pressures that favored the darker moths as being better adapted to the environment, not because they were parasites.

Ridley looks at this issue by asking the question, Why are we not more rational? His argument is to consider our ancestors a million years ago. Suppose that population consisted of two brain types: the Robots and the Irrationals. The Robots are governed by a rational control system, while the Irrationals hold to various religious superstitions to form their image of the world. Logically, natural selection should favor the Robots, since scientific rationality is more efficient than irrational superstition; it actually does transport people through the air, cures diseases, and provides labor-saving gadgets like TVs and dishwashers. In the struggle for existence, the rational protohumans would have reasoned about affairs objectively and done whatever was necessary to produce the best result. They would have built weapons whose efficiency was guaranteed by the laws of physics while their more spiritual contemporaries were conjuring with hocus-pocus. In a real conflict, the Robots would certainly have had the edge on the Irrationals. So there should be more Robot brains surviving in such an environment than Irrational ones.

As appealing as this argument seems, it may well be wrong, simply because it predicts the wrong winner. We are descended from the Irrational brains, not the Robots. Why? Well, no one really seems to know. But it seems that in a Darwinian struggle for existence, religious enthusiasm, for example, can crowd out rationality. Suppose, for instance, that the ancestral humans were a population of Robots. They calculate objectively how much effort to put into fights and other kinds of conflict and cooperative action, both inside and outside their group. They will do just fine—until they encounter mutants with a tendency to religious enthusiasm.

The religious newcomers value land not only for the resources it provides but because it has meaning to them: It is the home of their gods and must not be occupied by nonbelievers. So they take irrational risks in fights because the gods look after them in battle, and reward them later in the event of death on the battlefield. Some of the Irrationals may even exterminate their enemies when there is nothing rational to be gained by it for them. In this way, religious enthusiasm could crowd out rationality.

Note that the foregoing argument says nothing about the vulnerability of our own brain circuits to superstition. Nor does it appeal to the argument that religion causes us to identify with, or sacrifice ourselves to, the good of our local group. It simply follows the tenets of natural selection between two whole brain types in our past. Rationality alone is vulnerable because our reasons, and objective evidence, for almost everything are exceedingly poor. In the grander scheme of things, individual humans are almost irrelevant and we need something more than Robotic reason to prompt us to take action. So if we rely on reason alone, there is not much to stop us from saying, Why bother?

An Irrational, on the other hand, will suffer no such doubts. Natural selection will then favor people who go beyond the evidence, or believe more than they ought. So it is an important part of a religion that the gods should be interested in us. This puffs us up with self-importance, and makes us believe that we are the chosen creatures of supernatural entities who actually care about us. Any other kind of religion would be useless at crowding out rationality. So from an evolutionary point of view, if anything is a cultural virus it is science, not religion!

A Slot Machine, a Test Tube, and the Ghost of Lamarck

The cornerstone upon which the modern theory of evolution rests is that random changes in an organism's genetic makeup give rise to behaviors in the organism that are selected for or against by environmental pressures. Those organisms whose behaviors "work" in their environment are favored in the reproductive race, while those whose behaviors don't work are weeded out by the iron hand of nature. A key element in this picture is the assumption that the changes in the genetic makeup are random, and are in no way directed by the changing environment. In other words, changes in the genes and changes in the environment are completely independent. In the realm of molecular biology, this tenet has been enshrined as the Central Dogma of Molecular Biology, which states that information flows strictly from the genotype to the developed phenotype and never in the opposite direction.

In 1988, Harvard geneticists John Cairns, Julie Overbaugh, and Stephan Miller carried out experiments on bacteria strongly suggesting that under certain circumstances, information *does* flow in the opposite direction. Could Darwin have been wrong? asked many commentators when this work was announced. Was it possible that the French biologist Lamarck was right when he stated in 1809 that organisms could acquire behaviors favorable to their survival directly from the environment and have these characteristics pass directly into their genome? From a sociobiological point of view, this is exactly what happens in all the cultural adaptations we've discussed above. So if it could be seen in the physical structure of organisms as well, that would be strong evidence in favor of the idea that nurture is at least as powerful as nature in the determination of social behavior patterns.

The basic argument against the idea of Lamarckian evolution of the type reported by Cairns et al. stems from an experiment originally devised by Salvador Luria in the 1940s. Luria argued that if conventional evolutionary theory were correct and you grow a series of bacterial cultures in the laboratory, then most cultures should contain no mutants, some would contain a few mutants, and just occasionally there should be a culture very rich in mutants. Luria's rationale for this argument was the analogy of a slot machine in a casino, where most pulls of the handle generate no reward, a few give back a small gain, while only an extremely small fraction result in hitting the jackpot.

Jackpots in the bacterial cultures should be rare, because they depend on a chance mutation happening early in the life of a colony, since Darwinian theory allows for mutations only in the growing cells, not after they have reached their resting state. On the other hand, if mutations arise in direct response to environmental pressures, there would be no jackpots, since the number of bacteria challenged by the environmental agent would be roughly the same in each culture. The results to be expected for directed versus random mutations in four bacterial cultures are shown schematically in Figure 3.1. Part A of the figure shows the typical distribution of mutants (filled circles) under directed mutation, while part B shows what one might expect to see if the mutations are random.

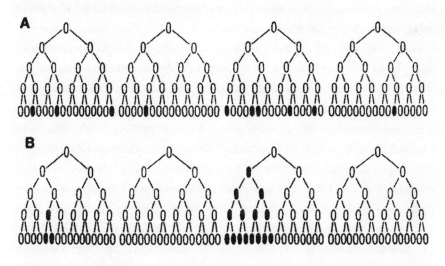

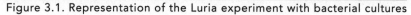

Figure 3.1. Representation of the Luria experiment with bacterial cultures

The results of Luria's experiment were completely consistent with the standard Darwinian party line: Most cultures contained few or no mutant bacteria, but a minority contained a large number. The implication was that genetic mutations do indeed occur at random during cell growth. Unfortunately, most biologists drew the far stronger—and unjustified—conclusion that this is the *only* route to spontaneous genetic change. And so things stood for over forty years until the Cairns experiments.

The Cairns group exploited earlier work by Francis Ryan, whose experiments indicated that there appeared to be a kind of "late" mutation in bacterial cells that were no longer growing. Those mutations arose in "resting" cells and were not the result of very slow cell division. Thus, there appeared to be two pathways to mutagenesis rather than one. The conventional pathway, identified by Luria, operates during cell growth and thus depends on DNA replication; the second pathway operates in resting cells, in the absence of measurable DNA replication.

The work of the Cairns group followed the path laid down by Ryan but added one crucial element. They asked the question: Do late mutations arise only under selective pressure, to solve the bacterium's growth difficulties, or do they happen regardless of such benefits? The answer from the Harvard experiments was that such mutations do indeed arise only under selective pressure. So not only do spontaneous mutations occur in the absence of conventional DNA replication, those mutations appear to be targeted so as to be advantageous to the bacterium.

The specter of Lamarck having thus been raised, defenders of the one, true Darwinian faith were quick to respond. Much of the ensuing debate, which rages to this day, centered on whether the late mutations really do arise in the absence of conventional DNA replication. Some suggestions had it that a limited amount of cell growth—and hence turnover of DNA—may have resulted from bacteria feeding on impurities in the culture medium, material from dead cells, or even nutrients produced by a tiny number of earlier mutants. But this concern was shown to be unfounded. A more important criticism was whether the unexpected mutations were truly Lamarckian. Here the focus of attention was upon the concern that the Harvard researchers had not looked at enough genes to be sure the mutations were specific to the gene under selection pressure. Later studies seemed to have laid this concern to rest. At present, the jury is still out on the issue of directed mutations, as arguments are presented almost daily about the inadequacies of controls in virtually every experiment that purports to show the existence of such mutations. Old prejudices die hard, especially when they involve such an icon of scientific faith as Darwin.

Since the crux of the argument in favor of genetic determination of behaviors rests on whether Darwinian theory can account for cooperative acts between organisms, let's finally turn to a careful consideration of this central issue.

The Selfish Altruist

In *The Prince*, his manual on how to wield power, Machiavelli comments on whether it is best to be loved or feared. After noting that it's most desirable to be both, but highly unlikely, Machiavelli comes down in favor of being feared. His reasoning is that "love is preserved by the link of obligation which, owing to the baseness of men, is broken at every opportunity for their advantage; but fear preserves you by a dread of punishment which never fails." In sociobiology this same principle applies in the realm of altruistic behavior.

Conventional sociobiological wisdom has it that cooperation often arises via the mechanism of what's termed *reciprocal altruism*. Basically, this means that if you help someone, especially someone closely related to you, then they are more likely than not to help you back at a later time. This is a kind of positive reciprocity, in which positive acts are repaid by similarly positive acts, thus leading to an evolutionary advantage to cooperation in a population. A large amount of the pro-sociobiological literature focuses on this mechanism to overcome the built-in bias of Darwinian evolution to favor selfish organisms over cooperators. A rather complete account of this mechanism is given in *Paradigms Lost,* so I'll refer the reader to that treatment for details. Here let's look at other mechanisms by which cooperation may occur when the positive aspect of reciprocal altruism is absent.

It turns out that there are at least six categories of interactions that can take place between unrelated members of a population. They are indicated in Figure 3.2, in which a minus sign represents a loss of fitness, a plus sign a fitness gain. Arrows show the initiator and recipient of the action. In punishing interactions, a selfish action by the first organism is followed by the first organism being punished by the second. Subsequently, the first organism modifies his/her behavior so that it increases the fitness of the punisher at some cost to itself. All six categories of interaction also occur between kin, though indirect benefits may alter the magnitude of fitness gains and losses.

First, let's focus on negative reciprocity, or punishment, in the context of evolutionary biology. Strangely, although retaliatory aggression is common in social animals, it has been largely ignored in the theoretical literature on sociobiology. Punishing strategies can be used to establish and maintain dominance relationships, to discourage parasites and cheats, to discipline offspring or prospective sexual partners and to maintain cooperative behavior. So it's

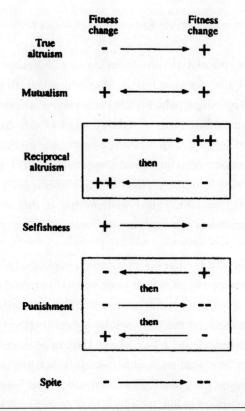

Figure 3.2. Six categories of interaction between nonkin

well worth examining how such a form of reciprocity influences the hypotheses underlying the genetic basis for behavior. We will return to the theoretical analyses supporting the existence of evolutionary stable strategies (those that persist under small perturbations in the gene pool) for negative reciprocity in the next section, but let's first make a small digression to consider one of the other forms of reciprocity, indirect reciprocity, which sheds some light on how moral systems evolve.

Conventional wisdom holds that cooperation and mutual aid require a tight partnership among individuals, or a close kinship relationship. But recently mathematicians Karl Sigmund of the University of Vienna and Martin Nowak of the Institute for Advanced Study in Princeton have shown that cooperation can become established even if recipients of the cooperation have no chance at all to return the help they receive to their helper. The reasoning underlying the Sigmund-Nowak result is that helping improves one's reputation, which in turn makes one more likely to be helped in the future by others.

This result differs radically from the conventional view because it assumes that any two organisms are unlikely to ever interact twice. Thus, direct reciprocity in which help would come from someone you had helped before is ruled out. Rather, it is indirect reciprocity that is at work here, in which your assistance is provided by somebody other than an individual that you have helped directly. For this type of reciprocity to work, individuals must assess others in their group by watching their interactions "from the side-lines," so to speak, and assign some measure of "image" to them. So if you watch your neighbor helping the little old lady across the street take out her garbage, then you might give him one point for image. On the other hand, if he refuses to help her when she asks for assistance, then you might subtract one point from your image of him. Then, when you are asked to help by watching your neighbor's dog when he goes on a weekend holiday, you will cooperate by taking his miserable, barking pooch if his score is high enough; otherwise, you "defect" by all of a sudden developing important business of your own that weekend. In this way, strategies for cooperation or defection can be more or less discriminating according to how high a score they require of the person requesting assistance before it is given.

But a new dilemma arises. When you cooperate, your only reward is a rise in your image score. This comes about at the expense of helping. You are betting that the benefit of helping will be reciprocated by someone who sees your high score and acknowledges your good reputation as a cooperator. On the other hand, your reputation gets tarnished if you punish a low-score player by refusing to help. This punishment is possibly unfair if, for example, the player you punish had the bad luck to encounter low-score players. In that case, you wouldn't help him because he had a low score himself as a result of not helping those others.

So to probe the success of discriminating strategies, Sigmund and Nowak simulated the effect of natural selection on a population that includes various levels of discrimination ("cut" levels). These levels range from blind cooperation (where every player one encounters receives help regardless of his reputation) to unconditional defection (where you never help anyone). In every generation of the simulation, there is a given—usually low—number of pairwise interactions, and the frequency of a strategy increases in proportion to its payoff. The overall conclusion from the simulation is that the best strategy is usually the cooperative one that discriminates the most. This means that its cut level is zero, meaning that you cooperate with others on your first interaction with them. In this way, cooperation is established even though repeated encounters almost never occur. These are mathematical results. Does anything even vaguely similar ever occur in nature?

The Arabian babbler is a gregarious type of bird, which seems to enjoy helping other babblers. Even more, the babblers actually seem to compete to see who can be the giver of help. The most natural explanation for this is that the increase in status compensates for the cost of cooperating. Similar phenomena are seen in the association of plants and various bird pollinators. While it's difficult to think of morality as an issue in the plant kingdom, such indirect reciprocity based on image scoring is a tempting explanation for some of the mutualistic tendencies plants appear to display. Such behavior again tends to favor the idea of genetic predisposition underpinning an important social behavioral pattern.

How do members of a population manage to discern cheaters and others of that ilk, in order to make effective use of indirect altruism? One way, many researchers now argue, is through the process of group selection, an idea that was espoused by Charles Darwin but fell into disfavor in the 1960s. More recently, researchers are active in bringing it back into the fold as an important component of the overall evolutionary process.

Return of the Group

The Hutterites are a Christian sect that first settled in the United States in the nineteenth century. They shun personal gain and stress the shared fate of the group, warning against the sin of selfishness. Anyone in the group who withholds help from others in need, turns lazy, or otherwise undermines community health and well-being receives a stern reprimand from church elders. Repeated offenses may even result in forced exile from the group.

The Hutterite leaders are elected democratically and undergo a long probationary period before acquiring full power. When a colony grows too large—which often happens, as the Hutterites have a very high birth rate—it splits into two groups of equal size, skill, and composition. A lottery then determines which group stays at the old location and which one moves on.

The beelike tactics of the Hutterites seem a lot like the process of cell division and evolution at the cellular level, suggesting to some researchers an evolved human capacity for thinking in groups and advancing group interests, even at the expense of personal aspirations. In short, groups like the Hutterites "can be functional units in their own right, and individuals sometimes behave more like organs in a larger organism than like organisms themselves," according to David Sloan Wilson of the State University of New York at Binghamton, one of the leaders of the charge to make group selection a respectable topic again in the world of evolutionary biology.

Group selection is simply the process by which natural selection affects groups of organisms, rather than acting at the level of an individual. So it involves the evolution of traits that boost the survival of some groups relative to others in a population. Before 1960, the concept of group selection was in vogue in the world of biologists. Researchers often spoke of social groups of animals and even entire ecosystems as harmonious units adapted to their surroundings. But in 1966, George C. Williams, an evolutionary biologist at the State University of New York at Stony Brook, published an epoch-making volume, *Adaptation and Natural Selection,* in which he blasted the group selectionist view as inaccurate and naive. Williams argued that genes engage in a titanic struggle for evolutionary survival, and that as they are passed down from one generation to another, they help individual organisms adapt to their physical and social environment.

Shortly after Williams's broadsides against group selection, mathematical models of kin selection and altruism, coupled with the emergence of evolutionary game theory (which we will consider in the next section), provided theoretical muscle for the opponents of group selection. The first of these theories suggested that altruism becomes more likely among genetic relatives, while the second offered a framework by which even genetic strangers can cooperate with each other through the vehicle of reciprocal altruism. But a new, reenergized version of group selection has started to emerge over the past five years or so.

Elliott Sober is a philosopher of science at the University of Wisconsin. He, along with Wilson, argues that natural selection preserves useful traits by means of a biological hierarchy that includes genes, individuals, groups, and populations containing interacting groups. According to this view, evolution is at work at *all* levels of the biological hierarchy. Thus traits can evolve that favor some genes over others in the same genome, some individuals over others in the same group, or some groups over others within a larger population.

Such a view of group selection has been criticized by some, who say that groups change their composition much too frequently to give natural selection long enough to work its magic. However, Wilson responds by saying that a group consists simply of a set of individuals influenced by the expression of an inherited trait, even if the group assembles intermittently and even if some of its members enter or leave at various times. Darwin, himself, took this point of view, arguing in *The Descent of Man* that human morality springs from this type of group selection.

In Wilson's view, cooperation and altruism evolve only by boosting the fortunes of one group versus another, while selfishness emerges through individual competition within groups. In fact, he maintains that kin selection and evolutionary game theory invoke their own forms of group selection to form alliances of individuals. For example, kin selection models depend on groups composed of individuals who have learned to recognize their genetic relatives, while game theory models involve groups composed of individuals who achieve cooperation by tracking the consequences of their interactions with others from the same population.

In this latter situation, Wilson argues, group selection may actually serve to *foster* altruism by the tendency of cooperative individuals to recognize one another quickly and accurately. Such a social *distant early warning* system allows cooperators to cluster together and leave predominantly selfish members of the population alone. So here we have a mechanism by which cheaters and others of low reputation or image can be recognized and effectively put into exile by the population at large. Simulations by Wilson and Lee A. Dugatkin of the University of Louisville have demonstrated this type of social sorting. These results are completely consistent with those of another study, by Robert H. Frank of Cornell University, who allowed groups of strangers to interact for a half hour. Group members were then able to predict with great accuracy which of their associates were most likely to act selfishly in later game-theory experiments.

While altruism and kin selection get lots of attention in the group selection debate, other phenomena as well enter into the picture. For example, in his influential 1966 book Williams argued that group selection should work to maximize a population's birth rate by endowing it with many more females than males. But if selection acts at the level of the individual, the ratio should be roughly equal. He then went on to claim that no known animal species exhibits the female-bias sex ratio that would follow from group selection.

Since that time, biologists have reported literally hundreds of species in which females slightly outnumber males (including our own *Homo sapiens*). Wilson suggests that this moderate, yet significant, preponderance of females reflects an ongoing tension between the opposing forces of group and individual selection.

Especially favorable areas for exploring the effects of group selection are those where pressures for large-scale cooperative action exist. In such situations, collective decision making assumes a vital role for some animals. For instance, Cornell University biologist Thomas Seeley characterizes honeybee hives as single "superorganisms," in which each hive renders decisions on a minute-by-minute basis on which flower patches to visit and which to ignore in an area of several square miles, how to allocate workers to either foraging or hive maintenance, and other matters of vital concern to the hive.

Seeley says that when the hive must make key judgments, each bee offers a small contribution to a chain of responses, which produces appropriate divisions of labor. At such times, hive members resemble neurons in a brain rather than independent agents. Similar types of decision making occur in prides of lions, where the group has to negotiate uncertain environments in which it pays to have the input of both adventuresome and cautious members. In all these cases, a case can be made for group selection as a vehicle for shaping collective decision making. But the battle is still far from over, as anthropologist Christopher Boehm of the University of Southern California notes when he says, "I know I'm attacking a cathedral of individual selection theory. . . . Advocating group selection as a force in human evolution has become like violating the incest taboo." And Williams, himself, has characterized Wilson by stating that he "engages in a kind of pedantic extremism by labeling all sorts of ephemeral groups as vehicles of natural selection." To this, Wilson replies, "Determined researchers can find ways to attribute cooperation, morality, and other group-oriented traits to the myriad deceptions of ultimately selfish individuals. . . . It will take decades for a full consensus to emerge."

It's All in the Game

In October 1994, three game theorists—John Nash, John Harsanyi, and Reinhard Selten—were awarded the Nobel Prize in economics for their work in game theory, a theoretical tool that untangles complex situations for finding the best strategy for a given player in interaction with many other players in a conflict situation. Game theory was originally developed in the late 1940s by the great Hungarian-American mathematician John von Neumann and the Austrian economist Oskar Morgenstern to explain markets and competition. But in more recent years, it has been used in other areas—including the study of animal behavior—to understand how organisms behave in the face of competition for scarce resources. Social scientists, for example, hope that understanding how a behavior such as cooperation is maintained in animals will yield insight into the origins of similar behavior in human systems. So in this regard game theory has much to teach us about the way gene pools may distribute themselves in a population. This, in turn, gives valuable insight into the way genetically based traits make their way through a population. Let's see why.

The big breakthrough in the use of game-theory ideas in biology came in 1973, when British evolutionary biologist John Maynard Smith used the ideas of game theory to explain when and why some animals, such as stags or fish, fight with each other. Maynard Smith proposed treating a given behavior as a strategy in a game and assumed that strategies evolve in just the same way as physical traits. Thus, any well-adapted population will follow the "best" strategy, in the sense that any mutants using a different strategy (behavior) will receive a lower reproductive payoff and will die out. Maynard Smith called this optimum strategy the *evolutionary stable strategy* (*ESS*).

Over the past two decades, theorists have studied nearly every type of animal behavior—aggression, cooperation, foraging, hunting, rivalry, you name it—by the theory of ESS. But to demonstrate that an animal really does follow an ESS, it's necessary for researchers to collect enough data to calculate the exact reproductive payoff for the observed strategy and its many alternatives. This effort has finally yielded the payoffs that biologists and mathematicians have been hoping for. So let's see one detailed example of this kind of analysis at work.

War-of-Attrition Game

A few years ago, Jim Marden of Penn State University and Jonathan Wage of Brown University collaborated in staging a series of territorial contests between male damselflies. The male pairs in these struggles had various, unequal reserves of fat, a measure of their strength. The common expectation for such contests (derived from game-theory models) is that each fly compares its own strength with that of its opponent and withdraws when it judges itself to be the likely loser. The duration of such contests is greatest when the opponents have nearly equal fighting ability, so that it's more difficult to judge the likely winner.

But duels between the male damselflies failed to follow this standard pattern. Although the weaker fly ultimately conceded to its opponent in more than 90 percent of the encounters, there was no negative correlation between the duration of the contest and the differences in relative strengths of the combatants. So here was a paradox. How did the damselfly contests differ from all the many other cases of animal conflict in which this negative correlation held?

One possible explanation is that damselflies don't assess their opponent's strength at all. In contrast to the earlier studies of other animals, in which the strength of an animal's opponent can be seen from the outside (for instance, in the size of the horns of bighorn sheep), in damselflies the fat is stored internally where it is difficult to observe by another damselfly. Moreover, experiments showed no correlation between fat reserves and physical attributes like wing span or body length that can be directly observed.

To investigate the hypothesis that damselflies know only their own strength and not that of their opponent, mathematician Michael Mesterton-Gibbons developed a game-theory model for such a situation. In the resultant game, known as a *war of attrition*, a strategy is based on the proportion of an animal's initial fat reserves that it is prepared to expend in a prolonged contest over a disputed site. The assumed payoff in this model is that the value of winning is proportional to the winner's remaining reserves. So the more fat reserves a damselfly has left, the more successful it will be in attracting a mate, finding food, and/or defending its territory in the future.

In the war-of-attrition game, the real ecological environment is reduced to two parameters, each a number between 0 and 1. These are the *coefficient of variation,* which measures the dispersion of energy reserves in the population, and the *cost-benefit ratio,* which compares the reproductive cost of a spent unit of fat reserves with the eventual winner's reproductive benefit from a saved unit. So one might expect that if the fat reserves are large enough, then an evolutionary stable strategy should always exist—regardless of the reproductive cost of spending a unit of reserve—since there's plenty of resource to waste. On the other hand, for smaller overall fat reserves in the population, a single unit becomes a scarcer and scarcer resource, and it may well be the case that no uniformly best strategy exists. The game-theory model tells us just where this crossover point resides.

In general, because an ESS is a strategy for the entire population of damselflies, such a strategy exists if the coefficient of variation exceeds a critical threshold. Specifically, the situation is as depicted in Figure 3.3. In mathematical terms, the curve dividing the region where the ESS exists from that where it does not is given by the expression

$$\text{ESS proportion} = \frac{1}{1 + \text{CBR} \times f(\text{CV})},$$

where CBR is the cost-benefit ratio and CV is the coefficient of variation. The function f increases from 0 to 1, so that the proportion exceeds $1/2$. If the point (CBR, CV) is not evolutionary stable, then any positive proportion of the population can be invaded by a mutant strategy adopting a proportion 0 (i.e., a strategy that gives up immediately in any contest).

At this ESS, both contestants are ready to expend at least half of their initial reserves on the contest, although only the loser actually does so. The reason this "no assessment" strategy is stable can be explained rather simply. Being prepared to expend too small a proportion of initial reserves may mean ceding needlessly to weaker opponents (who would otherwise lose), whereas too high a proportion may mean wasting reserves needlessly against much stronger opponents (who would win in any case). Although the damselflies don't know the reserves of their actual opponents, they achieve a balance between trade-offs by responding to the distribution of reserves among the entire population.

Other types of contests between animals have also been investigated using the same type of game-theory ideas, and the theoretical predictions of the ESS analysis has been borne out in practice. Readers interested in seeing

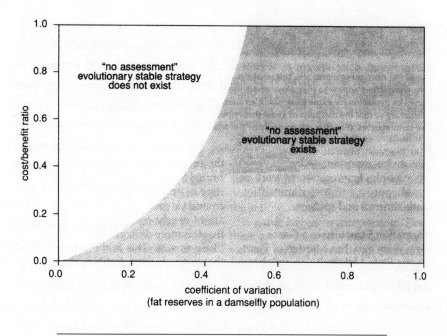

Figure 3.3. Existence region for ESS in war-of-attrition game

some of these results should consult the references cited in the "To Dig Deeper" section for this chapter at the end of the book.

To social scientists, the success of evolutionary game theory in offering a theoretical underpinning to cases of competition in animal populations is encouraging. But still more intriguing are cases in which game theory can explain why animals *cooperate*. A traditional game-theory model termed the Prisoner's Dilemma shows that cooperation is often not the favored strategy. Yet it has been shown by game-theory analysis that cooperative strategies can actually invade a population of egoists—under the right circumstances. Let's have a look at what those circumstances are in an attempt to understand how cooperation can come about naturally to counter Thomas Huxley's image of "Nature red in tooth and claw."

Prisoners, Cooperation, and Conflict

In its original form, the Prisoner's Dilemma game refers to a situation in which two prisoners are charged with a crime. If neither confesses—they "cooperate"—each will serve a minor sentence of, say, one year. If one cooperates by not confessing, while the other "defects," turning state's evidence against the first prisoner for the crime, the defector goes free while the cooperator serves a ten-year term. If both defect, both serve four years in jail. As they are interrogated separately, neither knows what action the other will take. So each reasons that no matter what his compatriot does, he'll serve less time if he defects (zero versus one year if his partner stays silent, four versus ten years if his partner sings to the cops). So both defect, thereby each getting four years of hard time—when they could have each gotten only one year if they had remained silent.

This kind of conflict between individual and collective rationality comes up all the time in human situations, ranging from business contracts to international environmental control agreements. Social scientists are bothered by the implication that the logical thing to do is to cheat. And, in fact, this strategy shows up in animals, too. For example, male bowerbirds build elaborate bowers—structures of twigs, leaves, and other objects—to attract females. But they also spend part of their time trying to damage and even destroy the bowers of other males. If all the male bowerbirds cooperated and left one another's bowers alone, all would benefit. But if any of the birds are marauders, they increase their chances for success with females at the expense of their nonmarauding rivals, making cooperation a losing proposition. But when we take a look at the world around us, we see lots of cooperation. So either this Prisoner's Dilemma situation is not a good way to picture conflict resolution in the real world, or there is something missing in the oversimplified game-theory structure that allows cooperative behavior to get a foothold in a population of cheaters. Most likely, both explanations contain some of the truth. But it's the second path that we want to explore now. How can we "soup up" the Prisoner's Dilemma game, so that cooperative behavior becomes evolutionarily advantageous?

Play It Again, Sam

The plain-vanilla, no-frills Prisoner's Dilemma game just described assumes that the two players meet just once, make their choice to either defect or cooperate, reap the fruits of their labors, and then go home. Of course, many situations in real life are like this—but most are not. Usually when we interact with someone, there is at least some chance that we may encounter that person again and be forced into playing the game another time. This is the so-called Iterated Prisoner's Dilemma game, in which after you play the game with a certain player, there is some nonzero likelihood that you will have to face that player again in another round of play. Under such circumstances, it has been shown by game theorists that cooperation can become a stable strategy.

A good example of such a game takes place among small fish. When a large fish nears a school of the smaller fish, one or more of the fish in the school will approach to see how dangerous the large fish is. This sort of "predator inspection" is risky for the scouts, but the information can be of benefit to them as well as to the rest of the school, since if the big fish is not a predator or is not hungry, the smaller fish don't have to scatter. According to Manfred Milinski of the University of Bern in Switzerland, a group of scouts approaching a big fish are playing out a Prisoner's Dilemma game: Each has a strong incentive to defect and let the others take all the risk. But if all defect, they learn nothing about the interloper. On the other hand, full cooperation minimizes the risks because the predator becomes confused if it can't focus on a single target. And since potential predators approach the school of small fish time and again, Milinski thought that a particularly robust strategy called Tit for Tat might have evolved among the small fish.

Tit for Tat is the simplest strategy imaginable. It is the game-theory analog of Golden Rule of "Do unto others as you would have them do unto you." The strategy says to always begin a round of play by cooperating. Thereafter, do whatever your opponent did on the previous round. So if your opponent tries to take advantage of your generosity by defecting on the first round, you defect back on the next round. And you continue in this mode until your opponent begins to be "nice" again by cooperating, at which time you "forgive" the earlier transgressions and begin to cooperate again yourself.

Milinski tested the idea that Tit for Tat evolved in sticklebacks, finding that the fish do indeed use a Tit for Tat strategy in predator inspection. Similar results were found by Lee Dugatkin of the University of Missouri, who looked at guppies. He says, "If one of them is trailing, the lead fish will turn around and head back. It will wait for the other to head out, and then it will go by its side." So if one fish defects (holds back), the other will too, and it then waits for the first one to cooperate (swim forward) before cooperating itself.

The guppies even remember from day to day what other guppies did. Dugatkin noted that it should be in the interest of guppies to associate with other cooperators, since it would be in their interest to be near cooperators if a predator appears. Experiments supported this conjecture, which suggests some new and interesting predictions about the evolution of cooperation in areas outside the guppy tank.

For instance, economists have a hard time explaining how markets end up in a Nash equilibrium, in which no individual can gain an advantage by unilaterally changing his or her strategy. Observations suggest that such equilibria do indeed arise—but the theory predicting them assumes that all the participants in the market are perfectly rational. This is certainly not supported by any evidence! Consequently, economists are looking to evolutionary game theory for processes other than rational decision making that could lead to a Nash equilibrium. Peter Hammerstein, a German game theorist, suggests that perhaps such strategies evolve in the same way that Tit for Tat evolved in guppies: by experience, not by rational calculation.

What all these results seem to be saying is that cooperation *can* emerge in a population of egoists—but only if there is some chance of the members of the population interacting more than once with other members of the group. But these are theoretical results, based on idealized assumptions and situations in the world of mathematical games. So, encouraging as they are in lending support to the idea that cooperative behavior can indeed be evolutionarily stable once it gets a start, the issue of how these game-theory conclusions relate to human behaviors in the real world remains open. Having come to the end of our story of how human social behavior patterns might be coded in the genes, let's take a moment to summarize the state of play.

THE APPEAL:
SUMMARY ARGUMENTS

Are criminals born or made? Is homosexuality a preference or a disposition? Do IQ tests measure innate abilities or acquired skills? These are the kinds of questions that pit biologists and sociologists against one another on opposite sides of the nature-versus-nurture divide. We have seen above that the evidence is growing that many such human social traits are indeed strongly, even dominantly, influenced by one's genetic makeup. Take criminality, for instance. In the 1980s, Harvard psychologist Richard Herrnstein and political scientist James Q. Wilson published a massive tome, *Crime and Human Nature*, in which they argued that the best explanation for a lot of predatory criminal behavior is probably biological rather than sociological. They then spoiled their case by appealing to "criminal types," saying that people with low verbal intelligence and *mesomorphic* bodies (short and muscular) are more likely to be criminals.

Such work strikes horror in sociologists, since it suggests an obvious remedy: selective breeding. In fact, Herrnstein suggests darkly that the greater fertility of stupid people means that the wrong kind of selective breeding is already taking place and may be responsible for falling academic standards. Other studies of this sort in areas ranging from the relationship between IQ levels and race to the correlation between head size and brain power all raise the hackles of the committed nurturist.

In recent years, a new assault on the nurturists has come from scientists calling themselves Darwinian psychologists. Their work argues that much sophisticated human behavior is not taught but develops autonomously, and that learning is not the opposite of instinct but is itself a highly directed instinct. A good example is Chomskyan linguistics, which claims that there is an innate language "organ" in the brain, specifically designed for learning language. If so, then the tendency to learn human language is human nature, not nurture, although the specific language that's learned is certainly dependent on the cultural environment into which the human child is placed. The Darwinian psychologists go one step further, however, saying if language acquisition is genetically based, why not other learned activities as well?

Sociologists and anthropologists cringe at such arguments, retreating in many cases to the position advocated by Émile Durkheim at the beginning of this century that the human mind is a tabula rasa—a blank slate upon which any cultural data can be written. One counterargument states that culture will explain human variation when there are reports of women war parties raiding villages to capture men as husbands. That will be some day!

To summarize our deliberations in this chapter, Table 3.3 outlines the various positions we've looked at for and against genetic imperialism. (Note: I have not repeated investigators' names in this table if they already appear for one side or the other in Tables 3.1 or 3.2.)

EVIDENCE	INVESTIGATOR(S)	EVIDENCE FAVORS
homosexuality	Hamer	Prosecution
social bonding in voles	Insel	Prosecution
twins' intelligence	Swedish-USA study	Prosecution
youngest ascendancy in macaque monkeys	Hill	Defense
gene-culture coevolution	Rose	Defense
memetics	Ridley	Defense
Lamarckism	Cairns	Defense
altruism(s)	Sigmund and Nowak	Prosecution
group selection	Wilson, Sober, Seeley	Prosecution
evolutionary games	Marden, Mesterton-Gibbons, Milinski, Dugatkin	Prosecution

Table 3.3. The evidence

THE DECISION: APPEAL UPHELD

Over the past decade, it seems to me, the evidence has mounted considerably supporting the claim that genetic makeup causes a predisposition to various types of human social behaviors. At the time of the writing of *Paradigms Lost,* the sociobiology debate really was a debate; today, the biologists are routing the nurturists, and the old Scottish verdict "Not Proven," rendered a decade ago, is simply no longer tenable.

FOUR

Born to Speak

Claim: Human Language Capacity Stems

from a Unique, Innate Property

of the Brain

BACKGROUND

The Linguistic Wars

Thomas Hobbes's 1651 novel *Leviathan* states, "When a man reasons, he does nothing else but conceive a sum total from addition of parcels, or conceive a remainder from subtraction of one sum from another. . . . For REASON is nothing but *reckoning*." This passage, in which symbols are represented as patterns of activity in the brain that trigger other patterns, became the basis for the rationalist philosophy of Descartes and Leibniz, and much later the information-processing models in cognitive psychology—including Noam Chomsky's theory of generative grammar. It is a compelling idea, but there is an equally compelling alternative.

In the 1748 treatise *Enquiry Concerning Human Understanding,* David Hume wrote that "there appear to be only three principles of connection among ideas, namely, *resemblance, contiguity* in time or place, and *cause or effect*" (emphasis added). Here Hume summarizes the theory of *associationism.* The mind connects things that are experienced together or that look alike, and generalizes to new objects according to their resemblance to known ones. This is the foundation of the *behavioral psychology* of John Watson, Ivan Pavlov, and B. F. Skinner.

To further show the difference between rational symbol processing and behaviorism, the two competing schools of thought in the problem of human language acquisition, let's turn to the tragic case of a thirteen-year-old girl given the pseudonym Genie.

Genie

In November of 1970, Children's Hospital in Los Angeles became the scene of one of the most fascinating—and bizarre—experiments in the annals of language acquisition studies. In that month, Genie came to live there. From the age of two years, she had been kept under restraints in a bedroom of a small house in the suburb of Temple City. Her mother, who was almost blind, finally managed to effect Genie's escape from this imprisonment, while her father-jailer committed suicide shortly after Genie's discovery.

Having lived eleven of her thirteen years in what was essentially solitary confinement, Genie was unable to speak when she arrived at the hospital. As a result, she quickly became an object of great interest for a passel of doctors and scientists, all of whom were interested in studying how Genie's social adjustment process would unfold. In particular, Susan Curtiss, a graduate student at UCLA in the area of language acquisition, was assigned to the team studying Genie.

Although Genie would be exploited by the research team in a number of ways that are beside our point here, one of several important facts did emerge from their studies: She didn't seem to be using the left hemisphere of her brain for language at all. When it came to its central function, Genie's left brain appeared to be functionally dead. Curtiss asked, "Why should this be so?" Her answer was that normal cerebral organization may depend on language development occurring at the appropriate time. But why should we have to learn something that many claim is innate? The response to this sensible query is that the brain organizes language learning, but that some stimulus is needed to organize the brain. So if Genie was any indication, our brains are physically formed by the influence of language through aural stimulus. The organization of our brain is as genetically determined and as automatic as breathing. But just as breathing is initiated by the slap of a midwife, so language is midwifed by grammar.

The investigation of Genie's language acquisition capabilities supports one of the two basic competing theories on the topic: Language is acquired by a special "organ" in the brain, and the process is set in motion by genetically programmed processes that direct the organization of the brain. This is the *innateness* or *nativist* view of language acquisition. The most prominent proponent of this view is the famed MIT linguist, Noam Chomsky, who introduced the idea in the 1950s. But there is a competing school, championing the view that language is a learned phenomenon, making use of the same learning mechanisms in the brain that we use when we learn to type or to tie our shoes. Let's briefly summarize the fundamentals of each position.

Language Is Innate

Boiled down to its essence, the nativist position on language acquisition is that there exists specialized neural circuitry in every human brain specifically designed for acquiring a child's native language. Chomsky called this language "organ" the *universal grammar*. It isn't composed of one kind of circuitry for, say, English, and another kind for Chinese, although it is designed for learning language and nothing else. The universal grammar contains many features that are not hard-wired in place at birth, but rather get fixed by the child's exposure to its linguistic environment. It's a bit like a TV set that has special circuitry for receiving television signals, and not radio, microwave, or any type of signal other than television. But there is a tuning knob on the set that can be turned to one or another of the various channels available. Each channel is analogous to a particular language like Spanish, Urdu, or Hebrew. Just as with a TV set, once the tuning knob is set by the viewer, the circuitry then shows that station and no other, in the same way when the parameters of the universal grammar are established by the linguistic environment, the child learns that particular language and no other.

One of the strongest arguments in favor of the nativist position is that once children have learned the basics of their native tongue, they are capable of generating an infinite number of linguistic statements that they have never heard before. The rationale is that if they have never heard these phrases and sentences before, how can they create them if language is learned simply by hearing statements and repeating them? This ability to generate statements they have never heard before is called the "poverty of the stimulus" problem. It suggests that the capacity for creating innovative linguistic statements is built into the universal grammar. So once the child sets the parameters of the universal grammar in order to learn a specific language, the capability is there for creation of an infinite spectrum of grammatically correct utterances.

Chomsky's theory is fundamentally about syntax; it's about the way words combine to form clauses and sentences. Since syntax involves only the shuffling about of *symbols,* like the letters of the Latin alphabet or the ideograms of languages like Chinese or Thai, it's perhaps not surprising that a fundamental aspect of Chomsky's theory involves what he called a *transformational grammar,* which constitutes the rules by which the two sentences "Bill broke the glass" and "The glass was broken by Bill" are transformed into each other. Chomsky posited that there indeed exists a set of such transformation rules, forming what he termed the *deep structure* of language. The *surface structure* is then manifested as the outcome of applying these rules to a basic linguistic unit like a phrase or sentence. So the two sentences above are equivalent surface structures of a single phrase involving "Bill," a "glass," and the verb "to break."

Many others, like Chomsky's MIT colleague Jerry Fodor, have added some bells and whistles to the picture of language acquisition residing in a specialized structure in the brain. But, in a nutshell, Chomsky's idea of the transformation rules embodied in the universal grammar is the pillar upon which rests the nativist view of language acquisition.

Language Is Learned

The opposing school of thought has several different branches, the most notable of which are the *cognitivists,* like Jean Piaget, who advocate language learning as a staged process involving internal mechanisms in the brain, and behaviorists, like B. F. Skinner, who regard language acquisition as no different in principle from the positive reinforcement procedures used to train pigeons to peck at certain levers in a cage if they receive food from them. Others supporting the cognitivist view include Benjamin Whorf and Edward Sapir, who argued that "the limits of my language are the limits of my world," as well as linguist Geoffrey Sampson, who supports the innateness of the hierarchical structure of language, but then departs from Chomsky's notion of how it is acquired. Let's take a little harder look at some of these arguments.

Skinner's idea of operant behavior is the psychological analog prevailing at the individual level of biological evolution at the level of the species. In Skinner's scheme, "good" behavior is reinforced, just as in nature, "good" mutations are selected. But in both cases, there is no reinforcement until after the action has taken place. Operant conditioning is designed to explain the emergence of novel behavioral patterns in the individual in the same way that natural selection explains the emergence of new traits in a species. In both cases, the role of the environment is more to select than either to reward or punish, although Skinnerian reinforcement can be likened to a reward, since it encourages continuation of certain types of behavior just as natural selection encourages certain types of mutations.

Given his predilections for seeing operant behavior in every corner of life, it should come as no surprise that Skinner devoted a considerable amount of his impressive reservoirs of intellectual and polemical energy to the problem of language learning. He was particularly interested in this question since he held the view that language and self-knowledge are intimately intertwined. In the Skinnerian view, all words are acquired on the basis of the *law of effect,* that is, by rewarding, ignoring, or correcting the performance of novices by more mature users of the language. As a result of the way the human brain is structured to learn, a child comes to identify its pet dog with a word such as "Spot," with this identification taking place through a sequence of positive reinforcements from parents and older friends who have developed a more mature use of the language. So in the

Skinnerian version of language acquisition, language is learned in exactly the same way (operant conditioning) and with exactly the same psychological mechanisms (unspecified) as the child learns any other skill, like typing, balancing a broom on its end, or telling time.

In the epistemology of Piaget the child does not come "hard-wired" to understand concepts, but has to create them as in his construction of the ideas of space, time, conservation, and so on. In this framework, the environment provides feedback about the quality of the mental structures the child creates; it does not simply imprint the right structures on the mind. Thus, for Piaget the world is not just "out there" waiting to impress itself on a blank slate. Intellectual development is a constant interplay between the child and his environment, with the child playing an active, structuring role. Moreover, the Piagetian sees all areas of mental development as being closely interconnected with each other. So as far as language acquisition goes, Piaget sees it as all of a piece with the other stages of intellectual growth, and he places no particular emphasis on language as opposed to the other skills the child learns. As a result, the Piaget school contends that the mind develops more as a whole across a spectrum of intellectual tasks than as a modular structure.

All of these positions—nativist and cognitivist, alike—have been explained in detail in Chapter 4 of *Paradigms Lost,* so we only summarize them here in Tables 4.1 and 4.2.

THE LANGUAGE DEVICE IS INNATE AND UNIQUE!

PROMOTER	ARGUMENT
Chomsky	universal, generative, transformational grammars
Fodor	modularity of mind

Table 4.1. Summary arguments for the Prosecution

LANGUAGE IS MAINLY LEARNING AND/OR NOT INNATE!

PROMOTER	ARGUMENT
Skinner	operant conditioning
Piaget	stages of cognitive development; interactionism
Sapir and Whorf	"language = world"; relativism
Montague	Montague grammar
Sampson	Popperian learning of hierarchical structures

Table 4.2. Summary arguments for the Defense

THE LOWER-COURT VERDICT

A decade ago the competing arguments in the language acquisition debate looked like the situation depicted above in Tables 4.1 and 4.2. At that time, the verdict handed down by the lower court was a resounding decision for the Prosecution, and the Chomskyites walked down the courtroom steps with broad smiles on their faces. Has anything happened in the past ten years to change this picture? It will be useful to begin our exploration of the current state of this debate with a brief look at how human communication via language differs from that of our animal ancestors.

THE APPEAL

On Human Communication

Laura Ann Petitto is a professor of psychology at McGill University in Montreal. She was also the primary teacher of Nim Chimpsky, a chimpanzee that was the subject of a pioneering study of language and animals carried on at Columbia University some years ago. The goal of this study was, in effect, to investigate the ability of a chimpanzee to acquire human language capabilities. In other words, is language really species specific, unique to humans?

Animal Languages

According to Petitto, the answer is clear-cut: Apes are very complex in terms of both their brain structures (cognitive capabilities) and communication modes. They can refer to objects, they can have intentions about things and actions, they can demonstrate many different cognitive activities. But there are key aspects of human language that they just cannot master. As Petitto put it, "No ape project or primate project . . . claims that these apes master all the aspects of human language. *Everybody* shows some humility about this!" (emphasis in original). She goes on to note that it is undeniable that other species communicate. But then cautions that communication and language are just not the same thing at all.

In the specific context of Nim Chimpsky, she claims that Nim was not attending to the stream of symbolic communication that is language; he never extracted the essential information flowing out and through the language stream. As she says, "These apes do not have lexical knowledge, vocabulary knowledge. They don't have phonetic inventory, the collection of basic speech-sounds or speech-forms, from which all human language is formed. If you don't know the basic building blocks of English, you cannot *make* English! You cannot engage in complex grammatical forms, talking about what you did yesterday, or what you're going to do tomorrow or next year" (emphasis in original). In short, the apes don't achieve complex syntax. Nor do they achieve referencing to abstract things that aren't physically present. Petitto conjectures the reason the apes don't do that is that they don't have the relevant brain tissue. As complex as their brains are, they just don't have the amount of complexity characteristic of the human brain itself or of its language regions. It's simply not there. Although the evidence is still missing to draw the same conclusion for whale and dolphin brains, preliminary studies suggest the same is true for them. So when it comes to communication *by language,* humans seem to be unique.

Let's briefly list just a few of the characteristics of human language that tend to separate us from the chimps and other animals. All human languages involve:

1. Formation of a large number of meaningful symbols (words) from a small set of basic sounds (phonemes)

2. Formation of an unlimited number of sentences by logically combining words using a finite number of grammatical rules

3. Use of sentences for socialized actions

4. The ability of any normal child to learn to speak the language

By way of contrast, no known system of animal communication shares all of these characteristics. For example, the dance of the honeybees doesn't involve symbols or sentences, nor is it learned. Similarly, chimpanzees like Nim don't form structured sentences. So when it comes to real language capabilities, humans seem to be the only game in town.

The Origin of Language

When considering the acquisition of language, it's also very tempting to think about the origin of language. After all, it's not unreasonable to assume that the mechanism by which one acquires language might well be influenced greatly by the way in which language arose in the first place. The problem is that no one has the foggiest idea of how language actually arose. In fact, in 1866 the Linguistic Society of Paris declared that the topic of how human speech had evolved had become so divisive that the society banned all discussion on the matter. And even today the evolutionary selection pressures that led to the development of language are hotly disputed.

Traditionally, research on the topic has focused on endocasts, the fossilized interiors of the skull, which allow investigators to compare the brains of primeval humans with those of their modern descendants. More recently, the natural history of the vocal tract has been a prime area of research, with the configuration and placement of the vocal tract in ancient skeletons being inferred from the shape of the base of the skull. These studies all aim to pinpoint the precise period in the development of *Homo sapiens* at which the relevant cognitive capacity and physical apparatus were in place to give rise to what today we could call language.

Anthropologists have also entered into the fray, speculating that the more complex a tool was, the more likely language was needed for its creation. By that criterion, the first evidence of language dates back at least fifty thousand years, to around the time people first migrated from Africa to Australia. The complexity of the boats needed to make such a journey speaks for an ability among people to communicate with one another through language. But what about in earlier times, when the most complex tools were simple spears and axes?

Recently, biological anthropologist Robin Dunbar has advanced what might be called the "grooming" theory of the origin of language. Dunbar first notes that the larger the social group, the more time the members of the group spend on grooming each other. He then discovered that the relative size of the neocortex, the seat of higher thought in the brain, also increases in proportion to the group size. His proposal, then, is that in large primate groups, those with large neocortexes, grooming may help maintain social cohesion. So why do humans spend almost no time grooming one another?

According to Dunbar's theory, this is exactly where language enters the picture. He suggests that speech evolved when human groups grew too large for grooming to be an adequate communal activity. Compared with social grooming, language is a far more efficient way of bonding, since several individuals can converse at once. Moreover, speech enables group members to gossip about other members who are not involved in the conversation. Sad to say, however, Dunbar waffles a bit when asked why natural selection would favor groups that speak over groups that merely groom. "Perhaps," he says, "humans became their own predators, and the larger the group, the better the chance for protection from other groups." Language would have made larger groups easier to maintain, and so the survival of those larger groups in battle with smaller ones would have favored the development of language. Maybe! Nevertheless, the grooming theory is intriguing, as it focuses interest on the study of the adaptive significance of language in social relations.

Now that we know that human languages differ substantially from those of any known animal communication system, and that language probably originated as an adaptive trait, we must ask ourselves, But what trait, specifically? And we don't really have a clue. So let's return now to the problem of human language acquisition faced by children today, not those of a hundred millennia ago. Before looking into the evidence, pro and con, on the Chomsky versus Piaget debate, let's take a short intermezzo to consider just how human infants go about acquiring the rudimentary skills needed to learn a language.

Out of the Minds of Babes

Alan Gardner of the University of Utah is one of the pioneers in the study of animal communication, having been the first person to investigate the possibility of chimpanzees having linguistic capabilities when he trained a chimp named Washoe to communicate using American Sign Language. Gardner notes that chimpanzees are silent animals most of the time. A group of ten wild chimpanzees of different ages feeding peacefully in a fig tree in Africa may make so little sound that an inexperienced observer passing below can fail to detect them. Not so with humans. Humans are a talkative lot, putting out around twenty-four thousand individual speech sounds in an hour of relaxed conversation.

We like to think we talk a lot because we have a lot on our minds. To dispel this fantasy, Robin Dunbar, author of the grooming theory of language origins discussed earlier, carried out a study of the topics of conversations in the senior commons room at prestigious British universities. He discovered that most conversations focus on such pedestrian and mundane matters as personal relationships and experiences, sports, leisure-time activities, and the like. Only 14 percent of discussions had anything to do with academic matters.

Infants, we are discovering, are great talkers—whether or not anyone is around to hear them. Over seventy years ago, Jean Piaget wrote about what he called "private speech" among children, and numerous books have been published on the topic since. But most theories of language development give scant attention to language as a medium for social and emotional interaction, focusing instead on its role in the exchange of information. But when we speak, the way we do it often gives clues about our age, social class, education, and region of birth, and our facial and vocal patterns transmit our attitude toward the person with whom we are speaking.

Recent thinking has it that it is probably the social and emotional aspect of language rather than the need to communicate information that motivates infants to talk. As John L. Locke, director of the Neurolinguistics Laboratory at the Massachusetts General Hospital, put it, "I think that in order to act like the people with whom they share an emotional relationship—people who spend most of their time talking—infants may draw the conclusion, on some level, that they too must talk."

If one takes the problem of discovering the biological substrate for the development of language and not just that for fully developed language, then the neurology of social and emotional communication must be examined along with the neurology of linguistic communication. We must look at things like how the brain recognizes faces or voices and the emotions they convey. What we know is that the left hemisphere of the brain is primarily responsible for the mechanisms for processing and producing speech. By way of contrast, right-hemisphere mechanisms process facial and vocal activity and determine the emotional signals embedded in this activity. It now seems evident that these two hemispherical activities could not have developed in isolation; they're like two sides of the same coin, and both are needed in order to determine the whole structure.

Current thinking has it that infants need to socialize and assimilate the behavior of others, and it gives them important personal information to convey before they have language and complex thoughts. The language mechanism in the infant brain—whatever that may actually be—consumes utterances that are stored in the infant mind through the actions of other, more socially engaged mechanisms. So the infant does not learn a language as an adult does by deliberately studying its grammar and vocabulary. Instead the infant learns language more or less accidentally, as a kind of side effect of communicating by nonverbal means. But how, exactly, does this occur?

John Locke has worked out a four-phase theory of language development for infants. His theory is not the standard milestone in the development of language, such as orienting to the sound of a voice, pointing, babbling, and the like. Rather, his phases are more fundamental stages in the development of linguistic capacity. Let's look at the basic features of each of the steps in Locke's theory.

• *Vocal learning:* Mother-infant attachment is the act that gets the linguistic process off and running. This bond is a channel over which emotion is routinely communicated by means of verbal and facial expression. As one might expect, the infant will not learn the language of strangers as well as the mother's tongue. Most of the vocal learning in the first year of life is perceptual; these sounds are mostly simple shrieks, coos, and cries. The infant then begins to produce well-formed syllables, which we interpret as babbling. Indeed, it appears that baby talk is composed mainly of words such as "mama," "dada," "bye-bye," which have the same repeating consonant-vowel form as babbling. Maybe this is because adults try to say things that babies can repeat!

It's interesting to observe that babbling is a consequence of biology rather than of the linguistic environment. When infants being reared in different linguistic communities are examined, we find that they have similar sound repertoires. If linguistic history were to be rolled back, it seems likely that we'd find "mama" among our ancestors' first words.

• *Storage:* Work by Elizabeth Bates at the University of California, San Diego, shows that infants comprehend words and phrases when they are as young as eight months, long before they are able to say anything and long before they could have learned the arbitrary rules and abstractions of grammatical language. The first recognizable infant speech may be a word or phrase that occurs frequently in the speech of others, such as "Where is it?" These phrases seem to be learned as a single unit by the infant, then stored away until the infant is actually ready to talk. This is reminiscent of studies carried out by Peter Marler at the University of California, Davis, on bird songs. Marler found that birds store songs for a long time before the learning is revealed by the birds' singing the songs.

Interestingly, the storage of phrases and words by infants seems to be carried out by the right cerebral hemisphere. Studies supporting this show that damage to the right hemisphere impairs language comprehension more than damage to the left hemisphere—if the damage occurs before the age of two. The right hemisphere's importance then begins to decline in the next stage of language development.

• *Analysis and computation:* At about the age of two years, a specific language module appears in the infant's brain. This module is analytical in development and computational in practice. It locates recurring elements in the infant's stored utterances, and thus learns the rules by which such utterances are put together by adult speakers.

The most compelling evidence in support of there being such an analyzer is the work by Marcus and Pinker at MIT. These studies show that sometime between the ages of 20 and 36 months children suddenly begin to make grammatical mistakes. Statements like "goed" for the past tense of "go" appear instead of the earlier—and correct—"went." These mistakes show that the children are now "computing" instead of simply reproducing what they heard earlier. On the surface, it might appear that the child is actually regressing in its language development. But they are really surging rapidly forward, having discovered the same rules that the rest of us use to form grammar structures like the past tense or the plural.

It's important to note, however, that this computational feat rests upon the success of prior feats of analysis, in which children discover that their stored utterances consist of smaller units of speech. Probably this analytical mechanism is forced into operation by limited storage capabilities for under-analyzed linguistic material. The problem is that new words and phrases are coming in fast and furious between the ages of 16 and 18 months, and unless the child discovers things like units of speech and its organizational principles, the new vocabulary would quickly outstrip available memory.

• *Integration and elaboration:* Once they integrate the analytical and storage mechanisms, they can quickly achieve a far larger vocabulary. Applied to stored vocal forms, structural analysis produces rules. Rules, in turn, impose organization on incoming utterances, expediting the learning of new words. For instance, knowledge that in English nouns are often preceded by "a" or "the" and that their plurals are often created by adding a final sound makes it easier to locate nouns in sentences and thus to learn new nouns.

At the same time that lexical capacity is expanding, sentence generation becomes more automatic. Fine-tuning of grammatical rules and memorization of irregular forms gives child speakers language that sounds more and more like that of the adults they are copying. Once this happens, children become *Homo loquens*—just like you and me, able to use words to convey thoughts to others and to communicate feelings that they would otherwise not be able to transmit.

An important component of Locke's theory of language development is the third phase, in which the idea of rules for processing the verbal input stored previously enters in a crucial way. But how do we know there are any such rules in the human language organ?

Stick to the Rules

Many cognitive scientists believe that the human mind is not naturally very good at the kind of reasoning based on following rules. They argue that rule use emerges later in childhood as a result of formal schooling and socially articulated rules, or as the result of extensive training that makes associative networks of neurons in the brain approximate rulelike behavior. It would certainly deal a serious blow to the nativist view of language acquisition if it could be shown that the brain does not use rules in its processing of linguistic inputs, since the very notions of universal and transformational grammars rest on rule-based principles coded into the human language organ.

Recently, work by G. F. Marcus and his collaborators has shown that infants as young as seven months can abstract simple rules from languagelike sounds, suggesting that rule formation is not a late add-on but there from the very beginning. The method used by Marcus is to present a stimulus repeatedly until the infants are bored, then present them either with stimuli of the same kind or of a different kind. The notions of "same kind" and "different kind" are in the mind of the beholder, so if infants pay attention longer to the different kind, visual or aural, they must be telling the two types apart.

In the actual experiments, infants were trained with "sentences" that follow one sequence, such as "ga ti ga" and "li na li" (an ABA pattern), and then were presented with sentences that contained different words and either the same sequence, such as "wo fe wo" (ABA), or a different sequence, such as "wo fe fe" (ABB). The babies listened longer to the "different" sequence, showing that they must have discriminated ABA from ABB, since everything else about the test sequences, such as actual syllables and their transition probabilities, was the same. Marcus also showed that a kind of associative network frequently touted as a model of language learning does not discriminate the patterns the way these infants do.

Nowhere is there any claim that infants lack an ability to form associations, that rule learning is uniquely human, or that the rule-learning mechanism at work in this experiment is the same one that babies use to acquire language later. But the demonstration suggests that the ability to recognize abstract patterns of stimuli is a basic ability of the human mind. So, in this sense, these results lend support to the Chomskyan notion that a rule-based mechanism is at work for human language acquisition. It's also of interest to note recent work by William Marslen-Wilson and Lorraine Tyler on how people process regular and irregular verbs.

The Regularity of Irregularity

Most verbs in English form the past tense by adding *ed*. For instance, *walk* becomes *walked*. When we encounter a new verb like *out-Jordan* or a nonsense verb like *zug*, we usually supply the regular past tense: *out-Jordaned* or *zugged*. These new forms couldn't have been memorized, so must have been created by a rule of the form, "add *ed* to the verb."

Regular verbs can be contrasted with irregular ones like *ring-rang* that form the past tense in a different way: irregularly. There are only about 180 such verbs in English—but they are among the most common verbs, and they are unpredictable as to how they form the past tense. So unpredictable, in fact, that their past tense is retrieved from memory, since it's hard to think of a rule that would work for generating it in even a majority of the cases.

Cognitive scientists using neural network models have contested this division into the word/rule categories, saying that their associative networks can learn such exceptions without using anything that even vaguely resembles a rule, and that therefore any qualitative distinction between regular and irregular verbs is a fiction. But evidence from several recent studies fits a key prediction of the word/rule theory, namely, that different parts of the brain are involved to a different extent when people process regular and irregular verbs.

This work, by people like Michael Ullman, found that patients with agrammatic aphasia (a difficulty in composing sentences usually caused by damage to the left anterior cortex of the brain) have more trouble with regular verbs than with irregulars. By way of contrast, patients with anomic aphasia (a difficulty in retrieving words, usually caused by damage to the left temporal and posterior cortices) have more trouble with regular verbs. Other studies using magnetic resonance techniques have shown that different regions of the brain are active for regular and irregular verbs.

All these results lead to the strong belief that regular and irregular verbs are neurologically distinct. This lends strong support to the word/rule theory, and a strong challenge to the non-rule-based ideas for regular verbs.

With all these notions of what infants do by way of creating language mechanisms, as well as the results on rule-based mechanisms in the brain, let's turn to the debate itself and look at some specific pieces of research that tend to support the Chomsky school's view that language really is innate.

It's All in the Brain

Christopher is an idiot savant whose remarkable linguistic skills support the idea of a language organ in the brain. He has been studied for several years by Dr. Neil Smith, a linguist at University College, London. According to Smith, Christopher is socially inept, avoids eye contact, and has a nonverbal IQ of 65. He cannot draw even simple figures, carry on any sort of lengthy conversation, or care for himself in the world of everyday affairs. Yet Christopher has learned sixteen languages and is a skilled translator. When Christopher met a speaker of Berber, which was a new language for him, he asked the Berber if the script could be written in Tifanagh, a medieval script used by Berber women for writing love poetry.

Tests reveal, however, that Christopher's language abilities are completely independent of his cognitive abilities. He never mulls over the meaning of passages and is unable to think about what he translates. This suggests that language is processed in a separate brain organ that can remain intact in a damaged brain like Christopher's. Conversely, when the language organ is damaged by stroke or other injury, certain aspects of language are permanently lost. An example is when English speakers with Broca's aphasia lose the ability to use articles of speech like *the, and, or,* and other connectives and function words as opposed to nouns and verbs. Let's look at some other recent work that also leans in the Chomskyan direction.

The Genetics of Language

We met Laura Petitto earlier in the context of her work with Nim Chimpsky and the possibility of animals acquiring humanlike language capabilities. After that work suggested that there really is something special about humans, Petitto began studying the question of whether our capacity for language is biologically controlled. She states her question even more strongly: "Is there anything about language that is genetically transmitted?" What she means by "genetic" is worth qualifying. It does not mean *determined,* as in a lead-pipe cinch, but rather a biological possibility. Whether that possibility becomes actualized depends on many factors: environment, context, and other non-genetic forces.

As her focus for addressing this overarching question, Petitto studied the process of infant babbling—in both normal children and in those born deaf. Briefly, what she discovered is that normally developing babies babble. Period. If they are deaf and their parents are deaf, and they are learning a sign language as their native tongue, they still babble—but with their hands. Petitto says this supports the observation that infants must be born with a sensitivity to certain patterns in their language input. It's not the method—sound or sight—that's important, but the patterns that the input contains.

As a result of these and other studies, Petitto proposes a new theory of how humans acquire language. It has three main components:

• *Pattern recognition:* Our brains must contain at birth some kind of pattern-recognition mechanism. This mechanism is sensitive to specific patterns that appear in the input the brain receives from the outside world. Petitto says that the brain has a dedicated architecture, specific clusters of neurons that are particularly sensitive to specific aspects of language structure, most importantly to rhythmic patterns carried by the language input, kind of like a computer's operating system.

One very important implication of this pattern-recognition ability of infants is that it allows them to begin acquiring language without yet knowing what language is. Possessing an innate ability to detect languagelike patterns allows babies to quickly detect the basic sounds that form words, and then to create words themselves without ever having any previous understanding of the meaning of the words.

• *Interaction:* The second component of Petitto's theory of language acquisition is that the pattern-recognition mechanism interacts with other brain processes. One such connection is between the pattern-recognition structures for languagelike sounds and those parts of the brain that govern different kinds of body movements. The result is a fine-tuned connection between a language's structure—including grammar and rhythms—and the motor production of these structures.

Another connection is between the brain's mechanism for detecting languagelike patterns and the parts of the brain that detect and process sensory information.

• *Environment:* The third leg of Petitto's language acquisition triangle is the environment. The sensory input from the outside world carries information the infant's brain uses as raw material to fashion the child's mother tongue. But Petitto says the exact nature of the environment is pretty much irrelevant. As long as the inputs the infant receives contain the relevant patterns, the infant will try to learn it. As she says, "It does not matter how that input gets to the brain's potential language regions. It can come through the eyes or the ears. What matters is the pattern that the input carries with it."

Petitto's three easy steps to language comprise a resounding vote in favor of the Chomsky position that there is a language organ in the brain specifically devoted to acquiring a child's native tongue. Here is some more evidence favoring this stance.

Talking Genes

Paul was a child in Montreal who by the age of three had not shown the slightest interest in talking. Even by the age of six, his speech was virtually unintelligible, a typical sentence being "Last time I'm waiting in the train station and there's another trains coming." But in every other respect besides language, Paul showed normal intelligence. What Myrna Gopnik of McGill University found when she studied Paul was that he readily grasped the mechanics of constructing a sentence with subject, verb, and object, but he just could not apply simple rules such as adding *s* to form plurals or *ed* to form past tenses. And even by the age of fifteen, after a decade of speech therapy, Paul still had difficulty with concepts that most four-year-olds pick up easily on their own.

In a television show in 1989, Gopnik saw the same speech anomalies displayed by an entire British family. The same phenomenon that Paul showed was present in a grandmother, four of her five children, and eleven of twenty-four grandchildren. After giving the family a battery of linguistic tests, Gopnik concluded that there is some kind of "gene" for language.

The affected family members seemed unable to understand even the point of a test for past tense usage. For example, when asked to complete the sentences "Every day he walks eight miles. Yesterday he . . . ," one adult responded, "had a rest." Even after being asked to use a form of the verb, the adult said "walk." Once corrected, however, the dysphasics were able to use a specific word correctly if it was required in later parts of the test. Gopnik theorized that these people can memorize past tense forms using *ed* but can't improvise and apply the rule to unfamiliar words.

These kinds of results are strongly reminiscent of the work described in the previous section about the formation of past tenses for regular and irregular verbs. There we saw evidence in support of there being completely different parts of the brain involved in these tasks. The Gopnik study of language-impaired people like Paul, and especially the British family, lends even stronger support to the case that such disorders have a genetic basis.

Corroboration for this view comes also from the study of patients with what's called Williams syndrome. Even though people suffering from this anomaly have severe cognitive deficits, they frequently have good command of language and an impressive vocabulary. For example, one teenager with Williams syndrome was asked to name all the animals he knew. His list included a yak, an ibex, and a koala. Yet such people may have difficulty in forming the past tense of irregular verbs, often applying the "add *ed*" rule to generate things like *catched* or *runned*.

This past tense disorder in Williams syndrome is just the opposite of that disorder shown by Gopnik's group: Some people with Williams syndrome have trouble forming irregular tenses but not regular ones, and Gopnik's group has the reverse problem. This suggests that in people with Williams syndrome the mechanism for retrieving words from memory does not function normally, whereas in Gopnik's group a separate mechanism for grammatical, or rule-based, speech has been impaired. Thus, since the impairment of one linguistic task does not automatically imply the impairment of other such tasks of the same relative degree of difficulty, it follows that language acquisition is not simply the by-product of a postbirth act of learning.

Finally, Williams syndrome also appears to be genetically linked, because it is caused by a deficiency in calcitonin, a hormone excreted by the thyroid gland and manufactured under the control of a single gene. Conclusion? Language is genetically based.

In this section we have seen some pretty strong evidence favoring the notion of a language organ serving to provide human infants with the biological potential for acquisition of language. But the opposition, those who feel that language acquisition is part of the general learning capability of the brain, have their arguments to put forth, too. Let's hear them out.

Chomping on Chomsky

Geoffrey Sampson is a linguist at the University of Leeds in the UK. A few years ago, he gathered together all the arguments employed by Chomsky in support of his innateness hypothesis, and produced a point-by-point refutation of them. Strangely, this set of arguments seems to have passed over the linguistic community, perhaps because it went against the temper of the times. But it's worth reviewing Sampson's case here, especially since his paper not only presents arguments against Chomsky, but also offers an alternative hypothesis for language acquisition based on the philosophy of learning of Karl Popper.

To do justice to Sampson's case, let's first review the major premises in Chomsky's theory of language acquisition.

1. *Speed of acquisition:* Children learn their native language remarkably fast, much faster than their acquisition of other bodies of knowledge, such as physics.

2. *Age-dependence:* Language acquisition in childhood is quite different from learning a language later in life. Adult language learning is rather slow, and its end result less successful.

3. *Poverty of data:* The child must infer the general rules underlying linguistic structures from individual examples of language that come to him or her during the acquisition period. Children are usually given little explicit instruction about the structure of their first language. Furthermore, the samples of language that the child is exposed to during the acquisition period is both small relative to the potential total sample, and qualitatively poor because of slips of the tongue, indistinct speech patterns, bad grammar, and so on by adults.

4. *Convergence among grammars:* Although children differ greatly in intelligence and are exposed to different finite samples of

their elders' language, almost all the children in a given language community acquire essentially the same language as one another and the same language that their elders speak.

5. *Linguistic universals:* All languages that are or have been actually used by humans resemble one another with respect to a number of structural features, which are by no means necessary properties of any conceivable "language."

6. *Nonlinguistic analogies:* Other human cognitive achievements that resemble language in being uniform across humankind and that are grossly underdetermined by experience are also used by Chomsky in support of his innateness hypothesis. For example, the capacity to deal with the number system, common to all humans (according to Chomsky).

These six premises are the bedrock upon which Chomsky's innateness hypothesis rests. Before discussing Sampson's arguments against them, let's first sketch the theory of language acquisition that Sampson suggests as an alternative to Chomsky's. It is based on the general learning principles espoused by Karl Popper in his "conjectures and refutations" philosophy of scientific knowledge.

According to Sampson, all the evidence we have about language acquisition supports the view that it is a process analogous to that of scientific theory formation, as envisaged by Popper. This is a process in which knowledge develops through creative formulation of hypotheses that are tested against experience. Popper suggests that the growth of linguistic knowledge in a child is parallel to the growth of explicit scientific knowledge, as this process is outlined in his hypothetico-deductive philosophy of science. But for Popper an innateness hypothesis like Chomsky's must in due course be discarded as having been superseded by the acquisition of actual language. What is fixed from birth onward is the urge to seek maximally strong and simple hypotheses that are logically compatible with experience, but not any aspect of the content of one's hypotheses.

So in the Popperian view, the individual mind begins life entirely ignorant of the nature of any language, and reacts to stimuli such as others' speech by making guesses about regularities underlying the data. These guesses are at first very simple and limited; many of them are negated by subsequent experience and are thus abandoned. But some are not negated and are retained. These are then built upon in the formation of gradually more sophisticated hypotheses.

With this very brief summary of how Popper would see the process of language acquisition, let's turn to Sampson's counterarguments against Chomsky, taking them in the order presented above.

SPEED OF ACQUISITION

There are two variants here: The speed of acquiring a language is *absolutely* fast and it is *relatively* fast. The first says that it takes a shorter time than one would expect given the complexity of the system to be acquired. The second asserts that language acquisition is relatively fast compared with acquisition of, say, knowledge of physics.

Sampson concedes that if knowledge of language is innate, then language acquisition will be completed sooner than if everything has to be learned from scratch. But the observed rate of language acquisition can't support Chomsky's claim of "absolutely fast" language acquisition until that claim can be made precise enough to yield concrete predictions for the rate of acquisition. Chomsky has never done this. Children take years from birth to learn language, not days, weeks, or even months. But why is two years, for instance, "remarkably fast"? How long would humans have to take to learn their native tongue for Chomsky to say that the speed of acquisition is no longer "fast"? Twenty years? Fifty years? So the notion of absolutely fast is vacuous.

As for the claim of "relatively fast" language acquisition, how can we compare the acquisition of the kind of tacit knowledge represented by a child conforming to a linguistic community with the conscious, explicit knowledge of the structure of a language that is not acquired universally or rapidly? The comparison with physics is a false analogy, since the tacit knowledge of physics involved in learning to pour a liquid without spilling it, to skip a rope, or to succeed in throwing a ball to where the child wants it to go appears to be learned at about the same rate as language.

AGE-DEPENDENCE

Here Chomsky claims that the human ability to acquire a first language diminishes sharply at a relatively early age, somewhere between the age of seven and puberty. With regard to *second* languages, this is certainly not the case. What appears to be true is exactly the opposite: Adults are actually better at learning a second language than are children, according to a recent study by Ellen Bialystok and Kenji Hakuta. But what about the studies showing that first-language acquisition does tail off sharply after puberty? Do those studies support Chomsky's innateness claim? No, says Sampson, because they fail to distinguish language learning from any other type of learning. According to the famed cognitive psychologist Jerome Bruner, "Any subject can be taught effectively in some intellectually honest form to any child at any stage of development." So much for the argument for age-dependence.

POVERTY OF DATA

The fact that a child has to derive general rules from particular examples in order to learn its mother tongue does nothing to support the Chomskyan as against the Popperian view of language acquisition. For Popper, all learning has this form. Rather, for Popper the problem is supposedly posed by the fact that the examples of language available to the child are meager and degenerate, in the sense that they are often grammatically incorrect and/or incomplete.

On the issue of degeneracy, Chomsky claims that the child lacking innate linguistic knowledge would have a difficult time working out the grammar of its elders if the data available included defective sentences. According to one estimate, if only 5 percent of the sentences the child hears are grammatically wrong, the problem will be significant because the sentences do not come with labels saying "ungrammatical."

Actual empirical investigations show, however, that the speech of mothers to children is very well formed. For example, one study found that only one utterance out of 1,500 spoken to the child was defective. One ungrammatical utterance out of 1,500 is surely a weak foundation upon which to build a conclusion as far-reaching as Chomsky's theory of innate knowledge, argues Sampson.

The premise about meagerness of input is even less well founded.

Assuming that a human language is a system comprising infinitely many potential well-formed utterances, then it's trivially true that any finite sample of utterances presented to a child is small compared with the total possibilities available. But the set of utterances that a child actually receives in the language-learning years can hardly be called "small" in any absolute sense. So Chomsky's point can certainly not be a numerical one. Rather, he seems to be saying that the child's sample will typically lack any evidence bearing on some particular properties of the language being learned, yet the system the child acquires incorporates these properties.

But the detailed facts about the child's input data scarcely matter to Sampson, since the case is logically inconsistent anyway. The obvious reply to anyone who asserts that a language has certain properties, no evidence of which is present in the data available to someone learning that language, is "How do you know?" So if data on some particular point of grammar were too rare to be encountered during language acquisition, there would appear to be no way that an adult could consciously ascertain that grammatical fact. So much, says Sampson, for poverty of the data.

CONVERGENCE AMONG GRAMMARS

This argument comes in two pieces: (1) different individuals' mastery of the grammar of their native language does not correlate with their level of general intelligence, and (2) even though different children are exposed to different finite samples of the language, they all come to acquire near identical grammars.

The first part of the argument was actually abandoned by Chomsky himself, who conceded that individuals who are more intelligent and/or educated than others have a greater degree of mastery of their common mother tongue. So the first strand of the convergence argument unravels.

The second strand can be answered by asking, How can it possibly be determined whether various individuals have formulated grammars that are more similar than they would have been had those individuals not been exposed to the same speech community? To do this, one would have to construct descriptions of speech of at least two members of the community, and then show that the two grammars agree on various points for which there is good reason to believe that the speakers would not both have been exposed to the same relevant evidence. So by the same reasoning that the poverty-of-data problem turned out to be logically suspect, we conclude the same for the argument about convergence.

LINGUISTIC UNIVERSALS

There exists an explanation for universals of grammatical structure that seems just as adequate an account of the facts as Chomsky's innateness, but which actually contradicts that hypothesis. This is Herbert Simon's observation that any complex structure that is built up gradually by trial and error from simple beginnings in an evolutionary way can be expected to have certain hierarchical structural features—irrespective of whether those features serve any advantage in terms of efficiency, fitness for the system's purpose, or the like. In other words, hierarchical structuring allows a system to be built up piecemeal rather than in a single continuous sequence. The reader will undoubtedly note the congeniality of this fact with the Popperian view of learning. But Chomsky's account of an individual's cognitive development is not at all evolutionary; language is already present at the outset, and only the details are filled in from experience. Yet the hard-core linguistic universals identified by Chomsky fit beautifully the properties that Simon's argument predicts will be found in any system built up by gradual evolutionary processes.

So we find that linguistic universals, perhaps the strongest pillar of the six that form the basis for Chomsky's theory, are in reality best explained by a theory sharply at odds with Chomsky's.

This concludes our all too brief summary of the main points in Sampson's critique of the Chomsky theory. There is much, much more available in the literature. So we refer the interested reader to consult Sampson's original paper, cited in the "To Dig Deeper" section for this chapter. Let us now move on to another witness in the appeal against Chomsky, this time one testifying to the notion that language itself, not a language organ, is the important evolutionary process.

Learnability

Christer Johansson is a computer scientist at Lund University in Sweden. He tackled the poverty-of-data problem by asking if an artificial neural network is capable of mimicking Swedish verb morphology by mapping a representation of the sound of the infinitive form to a representation of the sound of the past tense. In this way he hoped to shed light on the interaction between sound units, which in turn would allow us to see if the structure for verbs could be acquired using a device that has no rule-based structure to it at all.

Johansson notes that without negative evidence being presented to the child to tell which constructions are ungrammatical, there appears to be no solution to how to recover from overgeneralization.

What Johansson's experiments showed is that a neural network can indeed learn the relationship between the various sound units in Swedish verbs. Moreover, the network was even capable of generalization—it can make suggestions for past tense of verbs that have never been presented to it in its set of training data.

Johansson tentatively concludes that perhaps evolution has not created a language-specific device in our brains (a language organ); rather, language might have evolved so that it could be learned. All the elements required for language to evolve are present: distribution of items and structures, variation in these components, and selective forces. The selective force in language development is the ease of learning the structures of the language. The ease of learning is aided by frequency of occurrence and frequency of use. The irregular verbs in English and Swedish have a tendency to rank high in a frequency-ordered vocabulary.

So in Johansson's view the idea of the evolution of a language organ in the brain has the flaw that it is not clear what selective forces would be at work to select for this kind of device. People can—and do—live without grammatical language; for example, people who immigrate as adults. The transfer of important information is not dependent on grammaticality. However, the speed of communication in either direction might be. In this connection, Dunbar's grooming hypothesis for the origin of language discussed earlier is of interest. Maybe the use of language is more social and aesthetic than practical. Perhaps we use language for pleasure; it may simply fulfill our need to communicate our views and feelings, not information per se.

These ideas open up the issue of language evolving in such a way as to make it easy for a human-type brain to learn it. A natural question to pose at this point is what about a language that was not evolved along those lines at all, but evolved for a completely different purpose than for humans to fulfill our need to communicate? This would be the case, for example, if we were to ever have "first contact" with an alien intelligence. So, to conclude our study of language acquisition, the next section will take a look at a few possibilities of languages too hard for the human mind to encompass.

Too Hard to Handle

Easter Island, a tiny speck of land lost in the South Pacific, has always exerted a strange fascination for archaeologists and language scholars. While most famous for its gigantic stone carvings of human faces looking outward to the sea, Easter Island contains another mystery that has only recently been partially solved. This mystery surrounds the unique *rongorongo,* one of the world's few undeciphered scripts.

The script comprises parallel lines of characters, many of them bird symbols, hooks, and so on, engraved on wooden tablets. Figure 4.1 shows an example of the script from the Santiago staff, a five-pound wooden scepter acquired by the Chileans in 1870. This staff, which is about five feet in length, once belonged to an Easter Island leader, and is now housed in Santiago's Natural History Museum.

It seems that *rongorongo* was a very late phenomenon, directly inspired by the visit of the Spanish in 1770, when a written proclamation of annexation was offered to the chiefs and priests to be "signed in their native characters." As this was probably the natives' first experience of speech embodied in parallel lines, they adopted a method of script that employed motifs they had already been using in the rock art.

The script now survives only on twenty-five pieces of wood, scattered around the world's museums. Since the Peruvian slave raids of 1862 removed the last aristocratic or priestly islanders who could truly understand the tablets, their content has remained a mystery. The best guess of scholars who studied the script is that it is a rudimentary phonetic writing system, with picture symbols expressing ideas as well as objects. So the

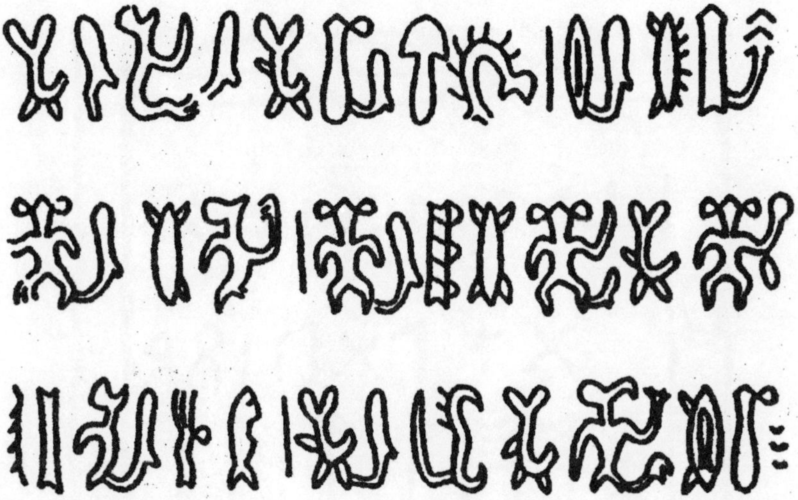

Figure 4.1. Detail of the *rongorongo* script from the Santiago staff

individual glyphs (symbols) did not represent an alphabet or even sylla-bles, but were "cues" for whole words or ideas, and a means of keeping count.

So what we have here is a written language that was produced by the same evolved brains that serve modern humans, but that has resisted all attempts to understand it—until 1996, when Steven Fischer of New Zealand developed a tentative reading. In Fischer's decipherment, the extract of script in Figure 4.1 says, "All the birds copulated with fish, and there issued forth the sun." The existence of scripts like *rongorongo* shows that even human languages can lose their ability to be learned when their native speakers die out, and calls into question the capacity of the human language organ to re-create/learn a written language that has no more "speakers."

Another example of this same sort is a puzzling inscription carved on a stone cross from the Dark Ages. What is left of the cross, which dates from the eighth or ninth century, is now in St. Peter's Church at Hackness near Scar-borough in North Yorkshire, England.

The cross carries five inscriptions. Three are in Latin and commemo-rate Abbess Aethelburg, probably of the local abbey. Until now, the other

Figure 4.2. The fifth inscription from the St. Peter's cross

two, written in strange alphabets, have been incomprehensible. One is made up of 27 letters and looks superficially like ogham, a script developed in Ireland in the fourth century. The fifth has 15 Anglo-Saxon runes, 35 tree runes, which look like pine trees, and 3 Latin letters. These are shown in Figure 4.2.

Recently, Richard Sermon, deputy director of the Scottish Urban Archaeological Trust, wrote a computer program to substitute known alphabets for the letters in the inscription in all possible combinations. He found that the Anglo-Saxon runes seem to be an anagram, which reads, "Aethelburg knew me." But the tree runes have left Sermon and his program stumped. There are up to four branches on the left and eight on the right of the "trunk," giving an alphabet of 32 letters, of which 14 appear on the cross. Sermon tried translating them into the 33 Anglo-Saxon runes, and wrote another program to generate all 24 possible versions. But, he says, "None appear to form any intelligible pattern. . . . It would seem that the tree runes are now too fragmentary to be fully understood."

So another script generated by the human brain's language organ passes into oblivion, unable to be understood by humans even just a thousand years later.

What does all this business about undecipherable scripts have to do with human language acquisition and the debate between the Chomskyans and the general learning theorists? Well, maybe not all that much other than to show that even languages that the human mind has created can become unlearnable by that same mind—provided the input environment changes dramatically enough. In these cases, the input environment essentially disappeared, leaving behind such fragmentary material for "training" that the language organ could not come to terms with it. In this same regard, the reader should contemplate the discussion of communication with extraterrestrials later in the book. Now it's time to turn our attention to a decision on the appeal.

THE APPEAL:
SUMMARY ARGUMENTS

The evidence has been flowing fast and furious on both sides of the language acquisition divide. Neurophysiological facts tending to support the existence of neuronal circuitry in the brain where the language organ seems to reside are pitted against logical inconsistencies in the Chomskyan picture of such an organ. And this is not to mention all the empirical evidence we have examined that leans in both directions at once. So to summarize our deliberations in this chapter, Table 4.3 outlines the various positions we've looked at for and against the idea that human language capacity stems from a unique, innate property of the human brain. (Note: As in earlier chapters, I have not repeated investigators' names in this table if their evidence already appears for one side or the other in Tables 4.1 or 4.2.)

EVIDENCE	INVESTIGATOR(S)	EVIDENCE FAVORS
chimpanzee language learning	Petitto	Prosecution
grooming	Dunbar	Prosecution
neurology of social and emotional communication	Locke, Bates, Marcus and Pinker	Prosecution
irregular vs. regular verbs	Ullman	Prosecution
aphasias	Smith	Prosecution
genetic disorders (e.g., Williams syndrome)	Petitto, Gopnik	Prosecution
logic and evidence	Sampson	Defense
learnability	Johansson	Defense
undecipherable scripts	Fischer, Sermon	??

Table 4.3. The evidence

THE DECISION: APPEAL DENIED

While Sampson's case against Chomsky looks appealing (no pun intended), the evidence he cites is simply not compelling enough to stand in the face of all the bits and pieces from so many different areas that favor the nativist position. And in the last decade there just doesn't seem to be all that much by way of dramatic evidence countering the Chomskyan view that there really is a language organ in our brains, which organ really is the seat of our ability to acquire and use language. So, if anything, the nativist case is even stronger today than when first examined in *Paradigms Lost* over a decade ago.

FIVE

Man-Made Minds

Claim: Digital Computers Can, in Principle,

Literally Think

BACKGROUND

Slaughter on Seventh Avenue

On the evening of May 11, 1997, human supremacy in the game of chess came to an abrupt end at the top of the Prudential Building in New York City when Deep Blue II, a computer program specially designed for chess by researchers at IBM, dethroned world champion Garry Kasparov. But what does this victory by Deep Blue II mean in the broader context of machine intelligence? Is it an indication that duplication—or even improvement—of full-blown human intelligence is possible in a computing machine? Or is it just a parlor trick, showing merely that a computer can play a world-class game of chess?

Interestingly, chess masters commenting on the Kasparov–Deep Blue struggle were puzzled by Kasparov's resignation in the fateful sixth game of the match. In the words of grand master John Fedorowicz, "Everybody was surprised that [Kasparov] resigned, because it didn't seem lost." Insight into the situation was provided by Kasparov, himself, who said after the game that he sensed a kind of "alien intelligence" in the machine. Donald Michie, one of the founders of artificial intelligence (AI), and a developer of one of the first computer chess-playing programs in the 1960s, wrote that for all its skill, Deep Blue has very little intelligence. He then added that he was in no doubt that Deep Blue came across to Kasparov as being able to read his mind, and that this slowly wore him down.

This latter remark by Michie underscores a large part of the difference between the way humans play chess and the way computers like Deep Blue II do it. In a given situation on the board, human players try to recognize overall patterns and structures of the pieces. They then explore a relatively small number of possible moves and countermoves from this position in deciding what piece to move next, correlating them with responses that have worked well in the past. By way of contrast, Deep Blue has very little pattern-recognizing capability at all. Rather, it has the ability to examine over 200 million possible moves each second from a given board position, and to evaluate the relative merit of each one of them. So the machine's strength lies in being able to explore this "tree" of possible moves, far beyond the depth possible by any human grand master. This radical difference in style of "thinking" about what to do in a given situation is no doubt responsible for Kasparov's collapse at the board in the final game of the match.

Deep Blue's speed means that it can make calculations during the time allocated to its opponent to consider and make a move, as well as during its own time. So if Kasparov took fifteen minutes to elaborate a plan and then moved accordingly, during this time Deep Blue has calculated several hundred billion possibilities. According to Deep Blue's programmers, the machine had predicted Kasparov's moves in about half the cases so it could reply instantly.

Now imagine the effect on Kasparov when a succession of his deeply pondered moves are answered instantly. When this happened during the tournament, he grimaced. Moreover, besides this spooky, "alien" effect, an instant response to a complex and difficult position gives rise to a peculiar form of time pressure not characteristic of human play. It allows no time for the human player to relax. And so to most knowledgeable observers, it was these psychological factors that ultimately led to Kasparov's undoing, not the brilliant moves of Deep Blue. It wasn't Deep Blue II that beat Kasparov; it was Kasparov.

The computer chess exercise illustrates nicely the difference between what AI researchers have called "strong" versus "weak" AI. Strong AI is the creation of a program that displays humanlike intelligence, solving problems the same way that humans do it (whatever *that* may be). Weak AI, on the other hand, is simply displaying humanlike intelligence by whatever means possible, which is the case with Deep Blue II in the context of chess playing. From the outside, all one sees is a set of decisions from Deep Blue. So an observer simply watching the match unfold by seeing the White and Black pieces being moved on a large board would not be able to distinguish Deep Blue from Kasparov. Such an observer might then be fooled into thinking that Deep Blue was simply a very clever human, who could just think a lot faster than Kasparov. And it would only be by looking "behind the scene," so to speak, and seeing that one player is a computer program and the other a human, that the observer could tell the difference between the two players. Such is the essence of the main criterion used by AI researchers to define intelligence, the so-called *Turing test*.

The Turing Test

In a seminal article in 1950 that sparked the development of the field we now term artificial intelligence, British mathematician Alan Turing suggested a criterion for deciding whether a machine was thinking. Basically, what Turing proposed was the commonsense notion that the way we decide whether other humans are thinking is by observing their behavior. If they display reactions to problem situations that look intelligent to other humans, then we deem them to be thinking "like us." For example, if I watch my son put on a pair of shoes and see him fasten the laces, then I assume he is thinking about the problem of how to keep the shoes secure to his feet and taking action accordingly. But if I put the same pair of shoes on my dog's feet, he does many things—bites the shoes, tugs at the laces, tries to kick off the shoes—but tying the laces is not one of them. And while I'm perfectly happy to believe that my dog is "thinking," he is certainly not thinking like a human.

In his original paper on the subject, Turing called this third-person view of intelligence the Imitation Game. Basically, it involves placing a human and a computer behind an opaque screen. The interrogator has a video display terminal that is connected to a similar terminal in front of the human, as well as to the machine. The interrogator then engages in a written dialogue with the machine and the human on the other side of the screen, being free to ask questions of any sort to either respondent. If after a sufficiently long period of such an interchange the interrogator is unable to reliably distinguish the human's responses from those of the machine, Turing says we must either say that the machine is thinking or that the human is not. This is the essence of the Turing test, and behavioral psychologists will recognize it as being firmly at the center of their view of cognition. Thinking is behavior—regardless of how that behavior is produced.

By one of those strange quirks of timing and fate that so often occur, at the very same time Deep Blue II was knocking Kasparov off his throne in New York City, just across town a competition was under way for the Loebner Prize, an award presented for the most humanlike conversation sustained by a machine. To make things interesting, the rules of the Loebner competition specify various categories of discourse, such as French wines, the nineteenth-century Russian novel, or baseball. Thus, what's involved is a restricted Turing test, since a full-fledged unrestricted test of the sort envisioned by Turing is manifestly beyond the capabilities of any existing program. As an amusing aside, the 1997 prize was won by Converse, a program developed by the British international chess master David Levy. This is the very same David Levy, who in 1970 made a bet that he would not be defeated by a chess-playing program for at least ten years. He won that bet, and managed to hold on for another decade until he was finally vanquished by a machine in 1990.

At first glance, the Turing test seems eminently sensible as a criterion for thinking, since it is basically the test that we humans employ dozens of times every day. But not so fast! There is a loyal opposition to the adequacy of this test. This opposition is embodied in a notorious—and ingenious—thought experiment devised by Berkeley philosopher John Searle. It goes under the rubric of the Chinese Room.

The Chinese Room

In August 1799, a Frenchman named Bouchard discovered a slab of black basalt near the Egyptian town of Rosetta, about thirty-five miles northeast of Alexandria. The stone contained inscriptions, apparently written by the priests of Memphis, summarizing benefactions conferred by Ptolemy V Epiphanes (205–180 B.C.); they were written in the ninth year of his reign in commemoration of his accession to the throne. Inscribed in two languages, Egyptian and Greek, and three writing systems, hieroglyphics, demotic script (a cursive form of Egyptian hieroglyphics), and the Greek alphabet, the stone provided a key to the translation of Egyptian hieroglyphic writing.

The decipherment of the Rosetta Stone in the 1820s was largely the work of Thomas Young of England and Jean-François Champollion of France. The hieroglyphic text on the stone contains six identical cartouches (oval figures enclosing hieroglyphs). Young deciphered the cartouche as the name of Ptolemy, thus proving a long-held assumption that the cartouches found in other inscriptions were the names of royalty. By examining the direction in which the bird and animal characters faced, Young also discovered the direction in which hieroglyphic signs were to be read (they can go either left or right, down, or, more rarely, up).

In 1821–22, Champollion, starting where Young left off, began to publish papers on the decipherment of hieratic and hieroglyphic writing based on study of the Rosetta Stone and eventually established an entire list of signs with their Greek equivalents. He was the first Egyptologist to realize that some of the signs were alphabetic, some syllabic, and some determinative, standing for the whole idea or object previously expressed. He also established that the hieroglyphic text of the Rosetta Stone was a translation from the Greek, not, as had been thought, the reverse.

Now suppose someone sits you down inside a closed room, in which there is a table, chair, and a largish book. Opening the book, you see two-column pages containing Greek characters in one column, with matching hieroglyphic symbols in the other column. You also notice that there is a slot in the door, into which someone (or something) from outside the room regularly sends in cards, each containing a Greek character. For example, you might receive the sequence of characters

ΑΛΕΧΑΝΔΡΟΣ

Looking up these characters in the book, you find the corresponding string of hieroglyphic symbols

As you understand not a single word of Greek or a single symbol of hieroglyphics, you fail to realize that this exchange of what are to you content-free symbol strings has resulted in the translation of the name "ALEXANDER" from Greek to hieroglyphics.

But you might argue that this is just a translation exercise of a particularly simple sort, and doesn't really involve the *meaning* of the term "Alexander" as a proper name at all. So to address this objection, suppose the entire exercise had been conducted in just one of these languages, say hieroglyphics. Imagine then that the symbol string sent is

with the corresponding output string

Even in this situation, you still have had no understanding at all of what these strings actually *mean*.

This is just the point raised by John Searle, who devised the foregoing closed-room thought experiment using Chinese ideograms (rather than Greek characters and hieroglyphic symbols). Searle's argument is that your use of the book to look up symbols corresponding to the input characters and subsequent outputting of the appropriate symbol is exactly what a computer program does in transforming a string of input symbols into a string of outputs. There is no understanding of any kind of the symbols by the person inside the room in this setup; hence, concludes Searle, a computer also has no understanding of what the symbols it is processing mean. And without such an understanding, Searle says, it's impossible to take seriously the idea that the machine is actually "thinking," in the sense that we use that term for human beings.

This "Chinese Room" argument really takes aim at the adequacy of the Turing test as a criterion for determining whether or not a computing machine is thinking. We'll come back to the arguments against the Chinese Room, as well as Searle's refinement of his case, later in the chapter. The main point to keep in mind here is that Searle is not necessarily arguing that a computer will never think. Rather, he's saying that if it does think it won't be by virtue of instantiating a computer program in the kind of syntax-only computing machines we use today. In Searle's own words in a 1989 letter to the author: "I say that syntax *by itself* is not constitutive of nor sufficient for semantics" (emphasis in the original).

While Searle's Chinese Room is a counterbalance to the Turing test, a much earlier argument against a machine thinking like you and me was put forward by the Oxford philosopher John Lucas. Rather than questioning the definition of what constitutes "thinking," Lucas put forth an argument intended to show that machines are subject to inherent logical limitations of a sort that do not impede human minds. Accepting Lucas's claim, one would then have to conclude that human minds are capable of doing something that a computing machine cannot; hence, a computing machine will never think like a human. The heart of the Lucas position rests on a famous result in mathematical logic proved by the Austrian logician Kurt Gödel in 1931. Since Turing actually anticipated this very objection in his pioneering 1950 paper, "On Computing Machinery and Intelligence," let's take a quick look at what Gödel actually proved—and why some people believe it has relevance to the question of strong AI.

Gödel's Theorem

The bedrock upon which the edifice of mathematics rests is the notion of proof. Unlike areas such as the law, where arguments can be won by force of personality alone, in mathematics an argument succeeds only by producing a logically consistent set of steps leading from a primitive axiom to the statement whose proof is desired. Such a set of deductive steps is called a *proof sequence,* with the final statement in the sequence termed a *theorem.* We're all familiar with such a setup from the elementary geometry of Euclid, which we grappled with in secondary school. In Euclid's view of the world, the axioms are "self-evident" truths, such as "Two points determine a straight line" and the infamous parallel postulate, "Through a given point, one and only one line may be drawn that is parallel to a given line." From a handful of such statements whose truth is accepted without proof, one can use the tools of deductive logic to derive a plenitude of theorems about the properties of triangles, circles, and other geometric objects.

In the early part of this century, the famed German mathematician David Hilbert believed that all of mathematics—arithmetic, geometry, analysis— could be framed within a logical system that would enable us to prove or disprove *any* statement you cared to make about mathematical objects. In other words, any assertion was either true or false, and which of these was the case could be determined in a finite number of deductive steps. Hilbert's dream of a unified framework within which to encompass all of mathematics was shattered in 1931, when the Austrian logician Kurt Gödel published a paper in which he showed that this could not possibly be the case. Gödel proved that for any consistent logical system strong enough to talk about the relationships between whole numbers (arithmetic), there must necessarily exist statements about numbers that could be neither proved nor disproved using the tools of that logical framework. Here by "consistency" we mean a system in which a statement and its negation cannot both be proved true, which is the minimal requirement for the system to be useful in separating fact from fiction.

Gödel actually proved even more. He showed that there must exist a statement about numbers that is unprovable within the rules of the logical system—but that can be seen to actually be true by looking at the statement from *outside* the system. Gödel accomplished this logical sleight-of-hand by inventing a clever way to code any statement about numbers using numbers themselves. He then coded the self-referential statement "This statement is unprovable" using his numerical scheme, thereby creating an assertion about numbers that is logically true—but cannot be seen to be true using the proof machinery of the logical system itself. One way to describe this "incompleteness" result is to say that Gödel proved that truth is bigger than proof.

So where is the connection between Gödel's Theorem and the problem of thinking machines? Basically, the connection comes from the fact that a logical proof sequence of the type Gödel says cannot prove every mathematical truth can be shown to be identical to the operation of a computer program. With a program, one begins with an input string of symbols presented to the computer (an axiom), and then follows the steps of the program (the logical operations on the axiom) to produce an output (the theorem). With this equivalence in hand, it's easy to see the argument anti-AI proponents raise against the possibility of a machine ever duplicating human thought processes.

The argument presented by Lucas in 1961 involves the following steps:

1. Gödel's result shows that there must exist a true statement about numbers that cannot be the end result of following a logical proof sequence.

2. Yet our human minds can see that this "Gödel statement" must be true.

3. A computer following a program is logically identical to a human following the proof sequence for a mathematical theorem.

4. Therefore, since the computer cannot prove the Gödel statement that our human minds can see to be true, we can know something that the computer cannot. Hence, it follows that the human mind transcends any possible computer program.

On the surface, this looks like a knockdown, airtight argument against the possibility of our ever creating a computer program that would demonstrate humanlike intelligence. But as we'll see later in the chapter, all is not as it seems. And there are plenty of counterarguments against this appeal to Gödel, not the least of which is that Gödel, himself, never believed that his theorem posed an insurmountable barrier to the existence of a thinking machine. We'll take up these counterarguments later in the Appeals section of the chapter.

There is yet a third line of attack against machines thinking like humans, the so-called phenomenological argument put forth in the 1960s by Searle's colleague in the Berkeley Philosophy Department, Hubert Dreyfus. The distilled essence of the Dreyfus position can be expressed in the following syllogism:

1. The AI community claims that thinking is just the manipulation of formal symbols according to a fixed set of rules for combining strings of symbols into more complex strings.

2. Phenomenology claims that knowing, understanding, perceiving, and the like involve more than just following rules; they also involve things like intuition and insight, which cannot be captured within any rule-based, symbol-processing framework.

3. Phenomenology is correct.

Therefore,

4. No amount of AI, however clever, will ever duplicate human thinking.

So Dreyfus, along with the philosopher Ludwig Wittgenstein, claims that there's more to life than simply following rules, since the rules don't contain the rules for their own application. It goes almost without saying, of course, that Dreyfus's detractors question every one of the premises on the foregoing list.

One of Dreyfus's favorite arguments involves the way in which one acquires expertise in the performance of some task like playing chess or driving a car. He notes that acquiring proficiency in such tasks requires one to go through several phases of learning, proceeding from the performance of a rank novice to that of a professional. And at each successive stage, fewer and fewer explicit rules are invoked to take actions appropriate for the situation at hand. In short, expert human knowledge involves much more than simply following rules. Or so goes the phenomenological claims.

With these anti-AI arguments on the table, let's move on to a brief consideration of the two main schools of pro-AI thought. They will show us how the creation of a thinking machine might actually be carried out, the so-called top-down and bottom-up approaches to strong AI.

Schools for Machines

Top-Down

Herbert Simon of Carnegie-Mellon University is one of the most respected American intellectuals of the last half-century. In addition to his pioneering work on decision making in organizations, for which he received the Nobel Prize in economics in 1978, Simon is one of the founders of what has come to be termed the top-down approach to artificial intelligence. This view of how to get a machine to think is exemplified by work Simon carried out with his late colleague at Carnegie-Mellon, Allen Newell, on creating a machine that could generate new mathematical results, a program called *Logic Theorist.*

The underlying principle employed by *Logic Theorist* and its successor, *General Problem Solver,* is a form of heuristic, or informal, reasoning called *means-end analysis.* What this involves is observing that when we have a problem to solve, we always start with a given initial state (data, premises, and other information about the problem), a set of particular states (goals) that we would like to reach, and a set of operations that can be used to transform one state into another. The task then becomes to find some sequence of operations that will transform the initial state into the terminal set. This all looks suspiciously like the procedures followed by a mathematician in proving a statement true by starting with an axiom and applying rules of logical inference to transform that axiom to the desired statement. And it is! So it's no surprise that Simon and Newell took this model from mathematics as the starting point for their approach to the creation of a thinking machine more than thirty years ago.

Simon and Newell supplied their programs with two kinds of heuristics:

• Procedures for detecting significant differences between two states

• Rules of thumb about which operators typically reduce differences between various kinds of states

The solution principle is then clear: Detect some difference between the initial state and the desired terminal state(s); apply some operator that ordinarily reduces such a difference; if the resulting state doesn't differ from the terminal states, stop; otherwise try the same procedure, but now from the new state reached, through application of the state-transforming operation.

Example: The Three-Coin Problem. To see how this kind of analysis works, consider the well-known Three-Coin Problem, in which we have three coins, each of whose initial position can be either heads (H) or tails (T). The goal is to transform the initial configuration into one for which all of the coins are showing either H or T; that is, the desired terminal states are HHH and TTT. For any given state, there are three possible operators: A, turn over the first coin; B, turn over the second coin; and C, turn over the third coin. A move corresponds to the choice of one of these operators, and a solution to the problem is a sequence of three moves that will transform the initial state into one of the two terminal states.

Figure 5.1 shows the sequence of possible moves that can be made in this game. Notice from the diagram that it is not possible to move from the state HTT to the goal state TTT in exactly three moves.

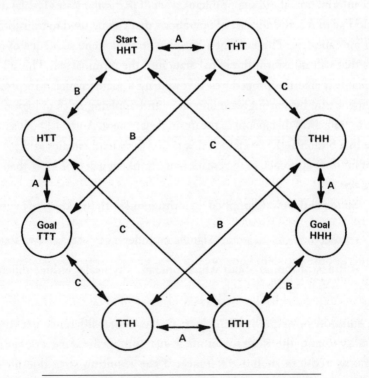

Figure 5.1. Possible moves in the Three-Coin game

The solving of logical puzzles, the playing of simple games like ticktack-toe, and a variety of other heuristic search activities typify what we term *automatic* formal systems. These are formal systems that work by themselves in the sense that in their normal mode of operation, they automatically manipulate the formal symbols of the system according to the system's rules. All of the Simon and Newell work on top-down computer cognition can be classified under the heading of such automatic formal systems. Unfortunately, several years' worth of experimenting with automatic formal systems has led to the unhappy conclusion that, rather than demonstrating that human thought is really just formal symbol manipulation in disguise, game playing, theorem proving, and the like can be done well without anything even approaching the full spectrum of human intelligence. In short, programs like *Logic Theorist* can produce intelligent-looking results in a very restricted domain, but once out of that domain there's a Grand-Canyon-sized gap separating them from what anyone would in polite company term "thinking."

The essential aspect of the top-down approach to strong AI is its view of thinking as tantamount to the formation of symbols and clusters of symbols in the brain, which are then combined by the machinery of the brain to form structures with semantic content. But nowhere does the explicit physical structure of the brain, its "architecture," so to speak, enter the picture. Top-downers believe that one can simply skim off the semantically laden patterns in the brain, ignoring the lower-level physical processes that actually give rise to these higher-level structures.

The problem of natural language-processing illustrates in the starkest possible terms the major difficulty with top-down approaches to AI. The programs just have no common sense. To illustrate this point, an early language translation program was asked to render the phrase "The spirit is strong, but the flesh is weak" into Russian. The resulting Russian phrase, when translated back into English came out as "The vodka was good, but the meat was rotten." Part of the problem here, of course, is that the word "spirit" can mean several different things, depending on the context. But the computer had no notion of context, which lies at the very heart of good language translation.

So we conclude that there's no way that programs of the top-down ilk are ever going to "think" in the way humans do until a way is found to code knowledge of the world into the formal symbols that the computer operates with. In retrospect, it's easy to see that when we perceive intelligently, we never perceive an object, but rather a *function* and a *context*. So if I show you a key, you never think of it as just a machined piece of metal; rather, you see it as an object that performs the function of unlocking something, perhaps a door, a safe, a car, or whatever the context suggests. It's this kind of knowledge that a computer needs if it's going to think—top-down style.

A number of disparate attempts have been made to deal with the commonsense acquisition problem for top-down AI. A couple of the more prominent efforts are the construction of microworlds and the concept of frames or scripts.

• *Microworlds.* A procrustean approach to giving computers common sense about the world is simply to fence off most of the outside world and let the computer have access only to a very severely restricted universe whose features, idiosyncrasies, folkways, and mores can be spelled out in painstaking detail and then given to the computer in some sort of digestible form. For example, the game Monopoly is a microworld in which the aspiring real-estate tycoons never have to worry about contingencies like fires, wars, deadbeat tenants, civil action suits, and the zillions of other annoyances that plague owners of pieces of real-world real estate. Board games like chess, go, and checkers are other microworlds of this type.

Probably the best-known microworld program is *SHRDLU,* a block world put together by Terry Winograd in the early 1970s. This universe consists of a few imaginary blocks of various sizes and shapes, strewn about on a flat surface. The blocks may be colored and cast shadows, but they never have any other physical properties beyond their geometric shapes and dimensions. *SHRDLU* knows all there is to know about this microscopic universe, and is able to converse in a seemingly intelligent fashion when queried about the world or asked to perform certain acts such as placing one block atop another, or picking up a block and moving it to a different location.

Despite what appear to be intelligent dialogues between *SHRDLU* and its inquisitors, the program has a variety of fatal deficiencies as a cognitive entity: (1) *SHRDLU* never initiates any actions but only reacts to queries put to it; (2) the program has no motivational goals whatsoever, other than the goals introduced by inquiries from the outside; (3) the main problems of perception and action involve capturing the interface between symbolic cognition and real objects. But *SHRDLU*'s "world" is *already* symbolic, so it doesn't address this interface at all. But these difficulties pale by comparison with the real problem concerning microworlds in general: They are capable of performing only because their domain is so stripped down that there is nothing left that could require even the slightest glimmer of understanding or real perception.

· *Frames and scripts:* These ways of giving a machine common sense are predicated upon the belief that few situations we encounter in daily life are really new. Technically, frames describe *static* situations, while scripts characterize a *dynamic* set of actions appropriate to a given set of circumstances. Most circumstances that we're called upon to deal with have enough in common with other situations that we can distill the principal features, analyze them, and store them for future retrieval and use. Thus, a frame acts like some of the questions one often sees on IQ tests, in which a scenario is created with a number of blanks left open to be filled in appropriately to demonstrate an understanding of the story. Although the frame idea appears to have originated with Marvin Minsky as an outgrowth of work on computer vision and language, the high priest of "frameology" is Roger Schank of Northwestern University, a somewhat controversial player in the AI game. What Schank's work demonstrates is that thinking and learning are not just passive processes of filing and retrieving information. The mind learns to build models and structures that can be continually modified and updated as new knowledge becomes available, and that dynamic knowledge base is used to plug the gaps in real-life scenarios as they unfold.

Of course, if a machine had only scripts it wouldn't be able to deal with novelty; it would understand only the prototypical situations that had been programmed into the scripts. Consequently, Schank and others, have been busy developing programs that would know about people's goals and desires, and how they might go about formulating plans to achieve them. One such program was tested on the story:

> John wanted money. He got a gun and walked into a liquor store. He told the owner he wanted money. The owner gave John the money and John left.

Nowhere does the story make mention of a robbery; nor does it explicitly state that the gun was used to threaten the liquor store owner. Nevertheless, the program was able to use its storehouse of knowledge about goals and plans in order to infer these facts.

Neither of these approaches—microworlds or frames—has turned out to be a panacea for the ills that plague the top-down approach to intelligent machines; nevertheless, top-downers continue to press on with their hopes of finally achieving the triumph of their rule-based, "symbol-crunching" style of AI, by putting the information into the program at the outset. We'll look at some of these efforts in a later section. But since it appears unlikely at present that further pounding away at such programs is going to get us substantially closer to a resolution of the basic thinking-machine question, let's move to the other end of the telescope for a look at bottom-up attacks on the matter of thoughts and machines.

Bottom-Up

According to most sources, the human brain contains around 10 billion neurons, connected in an impossibly dense network of axons and synapses. The neurons are the cellular equivalent of an on-off switch, and it is believed that all human cognitive activity arises from the switching on and off of the neurons in response to activation and inhibition signals they receive from other neurons to which they are connected. The structure of this neural network, along with the structure of the brain itself, is shown in Figures 5.2 and 5.3.

In light of the relatively meager results obtained by top-down researchers in over thirty years of effort, in the early 1980s workers in AI revisited the notion that perhaps there is something about the physical structure of the human brain that accounts for its cognitive capabilities; that perhaps the way to build a thinking machine is not to focus on symbols and their manipulation, but rather to construct a computing device whose physical architecture mimics as closely as possible that of a human brain. This "neural network"-type of machine might then develop intelligence in the same way that human children develop intelligence: by observing the world around them, and using these observations and the help of a teacher to learn. This is the essence of what's come to be termed the bottom-up school of strong AI, or what in some circles is called *connectionism*.

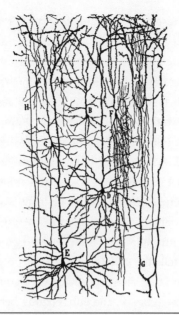

Figure 5.2. Neurons and their synaptic connections

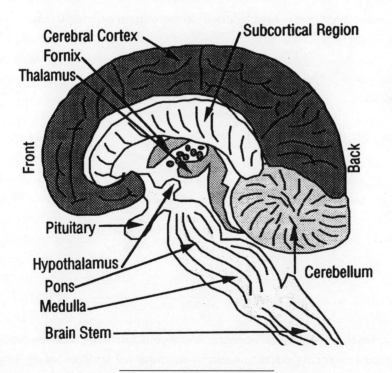

Figure 5.3. The human brain

The bottom-up research programs tend to fall into two camps distinguished by whether they focus mostly on software emulations of the brain or hardware mimicry of neural circuits. In the first group, we have researchers such as Douglas Hofstadter, Marvin Minsky, and neural network enthusiasts, in general. These folks make use of more or less standard serial computing machines to create software implementations of computational structures that mirror a primitive sort of brain. These neural networks have a structure like that shown in Figure 5.4, consisting of layers of mathematical neurons connected by channels that represent the axons and synapses in a real brain. A schematic diagram of one of these artificial neurons making up the network is shown in Figure 5.5. Signals are presented to the input layer of such a network, which percolate through the hidden layers to generate signals at the output layer. A "teacher" tells the system whether the output signal is correct, in some sense, information that then feeds back via a complicated procedure to change the numerical weights on the connections between the various neurons. This learning process repeats itself until a set of weights is found for the connections that produces acceptable behavior in the network. At this point, the neural net is said to have "learned" to recognize the input pattern.

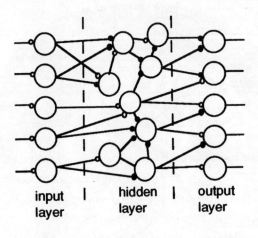

input   **hidden**   **output**
layer   **layer**   **layer**

Figure 5.4. An artificial neural network

On the other side of the connectionist coin are researchers who actually construct physical devices that mirror the neuronal structure of the brain. Such workers, like Gregory Hinton of Carnegie-Mellon University, in essence

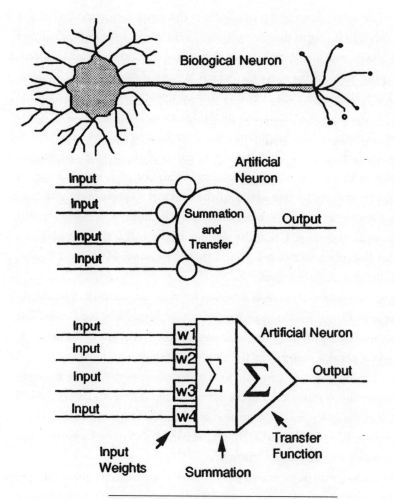

Figure 5.5. Diagram of an artificial neuron

are trying to construct an artificial brain based in silicon, metal, and plastic to simulate the "wet" brain we all carry around inside our heads. But the goal is still the same as that of the software-based connectionists: To get the machine to display behavior—usually pattern recognition of one sort or another—that passes for human. Here is a vastly oversimplified example of how this pattern-recognition process works.

A state of a neural network can be described as a point on a multi-dimensional surface, much as we can describe a point on a flat piece of paper by two numbers representing the point's x and y coordinates. For the neural network, there are as many coordinates as there are elements in the network. The value along each coordinate is the amount of activation of the element corresponding to that coordinate. So the state of the network is just a list of these coordinate values, a point in a multidimensional space. Changing the state of the network, which is how it "learns," can then be regarded as following a path over this surface toward lower computational "energy," in the same way that a ball on a slope rolls down to a state of lower gravitational energy. The weights on the links joining neurons in the network determine how the activity level of a neuron changes, so these weights determine the shape of this path. The learning algorithm is simply a way of adjusting the weights so that states of the network corresponding to patterns we want the network to remember are at the bottom of valleys in this multidimensional terrain. Using the analogy of the ball rolling around in a rugged landscape illustrates the notion that the system of neurons "rolls" down to the nearest stable state.

Figure 5.6 shows a situation in which we want the network to be able to recognize and distinguish the appearances of two dogs, Fido and Rover. The appearances of these two dogs are stable states in the landscape of the network. Imposing one of these patterns on the network is equivalent to placing the ball in a valley—it just stays put (diagram a). Adding noise to the system through, say, imperfect observations fed into the network's input neurons moves the ball up the wall of the valley, and if the noise is not too great it will roll back down to its old position (diagrams b and c). But if the ball is placed too far away it may roll down into the wrong valley (diagram d).

The analogy can be extended to show how a network can hold both specific and general information. For example, a limited number of distinct stable states can form in this surface (diagrams e and f). But if too many patterns are stored, some valleys may merge into one general valley (diagram g).

The bottom-up attitude of the connectionists sets their research program off from traditional AI in several ways:

• *Hardware counts:* It's simply not possible to separate the message from the medium; high-level symbolic processing at the level of the overall appearance of Fido and Rover cannot be abstracted from the lower-level hardware, like the neurons in the network. In the same way, measuring how your neurons are firing doesn't enable you to discern the neuronal firing pattern corresponding to your grandmother's face. Note that this does not imply that

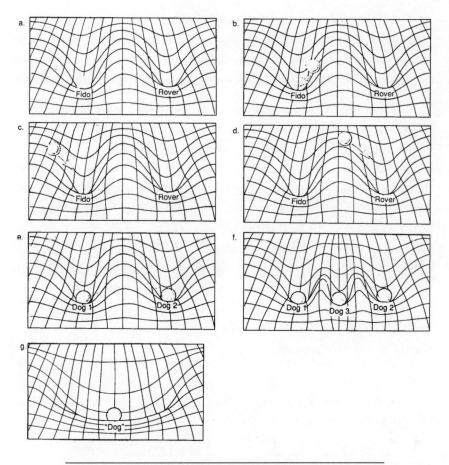

Figure 5.6. Multidimensional diagrams of a neural network

all such processing, and hence thinking, must be carried out in a medium like the human brain—only that hardware constraints matter when it comes to consideration of the cognitive powers of such processing objects.

• *Parallel architectures:* Connectionist computation is carried out in massively parallel machines, so that lots of computations are going on simultaneously.

• *Distributed processing:* Connectionist machines are deliberately diffuse in their memory and processing; the activities are spread around among the various processors with no single supervisory controller having overall command.

• *Unprogrammed:* The most striking feature of connectionist machines is the relative lack of specific instructions. Rather, there are a few general instructions, with the network finding its own solutions by settling down into stable states instead of following detailed, prespecified algorithms.

Objections to the connectionist view of cognition come in two flavors: theoretical and practical. On the side of theory, the biggest difficulty is that connectionism offers no clear-cut procedure for getting from the low-level energy states of neurons firing to high-level symbolic processing involving things like the overall appearance of Rover and Fido; that is, there's no prescription for bridging the gap between computation at the hardware level and actual cognition at the level of the software. Critics readily agree that when you turn a connection machine on, something is likely to emerge. But it's not likely to be thinking. The practical objection is that intelligent thinking can never be done in a connectionist network because you could never build a machine with enough connections. Connectionists reply that beyond some minimal level of connectivity, it may be possible to substitute faster switching speeds for more connections.

YES, COMPUTERS CAN THINK!

PROMOTER	RESEARCH PROGRAM
Top-down School	
Turing	Imitation Game
Simon and Newell	rule based, symbol manipulation
Schank, Minsky	script following, frames
PROMOTER	RESEARCH PROGRAM
Bottom-up School	
Hofstadter	subcognitive modules
Minsky	"society of mind"
Hinton	Boltzmann machine, statistical mechanics

Table 5.1. Summary arguments for the Prosecution

THE LOWER-COURT VERDICT

As of a decade ago, a telegraphic summary of the competing positions for and against humanlike computer intelligence would look like Tables 5.1 and 5.2, which indicate the competing arguments along with some names of prominent researchers identified with the various positions.

NO, COMPUTERS CANNOT THINK!

PROMOTER	ARGUMENT
Searle	Chinese Room
Dreyfus	phenomenology
Lucas	Gödel's theorems

Table 5.2. Summary arguments for the Defense

In 1989, the relative merits of these competing arguments led to a verdict in favor of the Prosecution's case that computing machines could, in principle, think like humans. At that time, the most convincing case seemed to come from the work of the bottom-up researchers, who were gearing up substantive strong-AI research programs to exploit the ever-improving computational technology just then starting to become available. Since that time, a lot has happened in the world of computing, cognitive science, philosophy of mind, neurophysiology, and many other areas bearing on the thinking-machine question. So let's revisit the whole issue by way of an Appeal to a higher court of this 1989 verdict.

THE APPEAL:
ARGUMENTS FOR THE PROSECUTION

The Creative Computer

On November 8, 1991, the Boston Computer Museum held the world's first hands-on Turing test, in which eight programs conversed with human inquisitors on a restricted range of topics that included women's clothing, romantic relationships, and Burgundy wine. This competition was spurred on by a $100,000 prize offered by philanthropist Hugh Loebner for the first program that could pass an *unrestricted* Turing test, as opposed to the restricted versions of the Boston competition. At the day's end, the judges awarded first prize to a program called *PC Therapist III,* which was designed to engage its questioner in a whimsical conversation about nothing in particular. For example, at one point the program suggested to a judge, "Perhaps you're not getting enough affection from your partner in the relationship." The judge replied, "What are the key elements that are important in relationships in order to prevent conflict or problems?" "I think you don't think I think," responded the machine.

As a result of the vast distance between this first set of entrants and any program that might even possibly pass the unrestricted Turing test, two subsequent Loebner competitions have been held at which Loebner opened his checkbook just a little bit, offering a $2,000 prize for the program judged "most human." The best results in all these tests have tended to come from programs employing tricks, such as deliberately mistyping replies, creating the illusion of "listening" by repeating snatches of an interrogator's input back, and steering the conversation the computer's way when it is in trouble by introducing a humorous remark so as to make the machine seem more human.

But these sorts of interchanges did little to fool the judges, most of whom said they were able to spot the mistakes rooted in a lack of everyday common sense, which immediately singled out the computer programs from the humans. Nevertheless, the overall conclusion from the Loebner Prize experiments is that perhaps the Turing test isn't as difficult as many people originally thought, since even the primitive programs in this contest managed to fool some of the judges most of the time. Of course, we have to bear in mind that these were not the true Turing test, since the domains of discourse were severely restricted. But it was still a pretty good start.

A telling argument against the adequacy of the Turing test as a benchmark of intelligence has been advanced by philosopher Ned Block. Suppose, he argues, that we write down a tree structure in which every possible conversation of less than five hours' duration is explicitly mapped out. This structure would clearly be enormous, much larger than any existing computer could store. But for the sake of argument, let's ignore this difficulty.

By following this tree structure, the machine would interact with its interrogator in a way indistinguishable from the way an intelligent human being would do so. Yet the machine would simply make its way through this tree, which strongly suggests that the machine has no mental states at all. And this same conclusion holds for any conversation of finite duration.

From this line of argument, Block draws the moral that thinking is not fully captured by the Turing test. What's wrong with the tree structure is not the behaviors it produces, but the *way* it produces them. Intelligence is not just the ability to answer questions in a manner indistinguishable from that of an intelligent person; to call a behavior intelligent is to make a statement about how that behavior is produced.

Racter and BACON

To hammer home the point that intelligence involves more than simply acting intelligent, let's consider the details of another computer-human interactive dialogue, the well-chronicled case of *Racter,* the computer raconteur. Here is a snippet of dialogue between *Racter* (R) and an artist, Joan Hall (J).

R: Would you please speedily yet cleverly tell me your name?

J: Joan Hall.

R: Shall we hastily or leisurely talk a little?

J: No.

R: Now, Hall, aside from being loving, what are you, and from where?

J: An artist from New York.

R: Joan, in view of the disgusting fact that you're an artist, listen to this. Years and days are spooky, Joan, now that my dissertation is insane. Some cosmologists from New York were singing to some other cosmologists. "Anyway our commitments are insane." Briskly the first cosmologist sang. By the same token the ambiguity is important, at all events, Joan, such tales are captivating. Do you not agree?

J: I suppose so.

R: You cried "I suppose so." Anyhow captivating cosmologists will never be arrogant. "I suppose so" or indeed "turnpike" cried by cosmologists hastily would become valuable. Suddenly they prepare for a commitment; yet enthralling hours are as minutes. In a millisecond we will get ready for enrapturing and arrogant dissertations. Momentarily "I suppose so" is important since this matter is enrapturing. You are exhausted, nevertheless an artist will speculate about neutrinos. Are these your views?

J: My view is that you are out of your mind.

R: Now wait. You sang "I suppose so," you now yodel "my view is that you are out of your mind"; drunkenly and cunningly I have now glimpsed your angry anguish. You, an artist, yodel "my view is that you are out of your mind."

Okay, I think you get the idea. A conversation with *Racter* is eerily similar to talking with a schizophrenic. *Racter*'s ramblings seem to constitute a coherent unit of conversation, but one that wanders aimlessly from topic to topic with no overall organizing center or any apparent goal. However, it is difficult to pinpoint where, exactly, *Racter*'s replies differ from those of a highly excitable patient in a psychiatric hospital. All this from a simple program relying upon nothing more than an endless sequence of iterative applications of a few simple rules. But what, exactly, are those rules? Unfortunately, it would take us too far afield to go into a detailed discussion of *Racter*'s innards. But we can take a quick look at a couple of the mirrors behind *Racter*'s magic.

Racter begins by selecting an item at random from a file. If the item is a literal statement, like "Tell me more," *Racter* simply prints it out directly. But it's far more probable that *Racter*'s random choice will yield a command, which will send the program off into other files that may themselves contain commands. When the initial command has finally been executed, *Racter* goes back to its files and randomly selects another element, thereby beginning a new cycle.

In order to begin a sentence, *Racter* must choose a form for the sentence, either at random or by some other rule that takes into account the program's recent dialogue. And in order to give its sentences some punch, *Racter* uses various identifiers, which are two-letter labels attached to words that are associated with different words. These identifiers cause *Racter* to make associations between successive words and sentences. As an example, *Racter* might come up with a sentence template like

THE noun.an verb.3p.et THE noun.fd.

where "an" is an identifier for *animals,* while "et" is associated with *eating* and "fd" is a tag for *food.* This template constitutes a rule for sentence generation. The program first searches its files for a noun, but only those nouns bearing the "an" tag. Thus, it would choose at random among nouns like *lion, tiger,* and *cat.* Next, having selected an animal, for instance *tiger, Racter* chooses a random verb bearing the identifier "et." These verbs involve eating and might include such words as *eat, chew,* and *chomp.* Again, a random choice from among this set might lead to *chew.* Having chosen the verb, the program next forms the third-person past tense indicated by the code "3p" in the sentence template structure. Finally, the directions tell *Racter* to pick a noun carrying the food tag, "fd." This may lead to the word *trout.* Putting all these choices together, the program spits out the sentence

The tiger chewed the trout.

Before leaving *Racter,* it's worth noting that the program can do much more than just select words with identifiers. It can generate its own sentential forms and even its own command strings, inserting them into the conversation at will. The interested reader can find out more about *Racter* by consulting the material cited in the references for this chapter. The examples of generating schizophrenic dialogues and conversations about Burgundy wines show the power computing machines have to fabricate new kinds of structures that have never existed before, a kind of creativity if you like. And these sorts of interactive discussions are not the only types of programs that top-down researchers have written that demonstrate some measure of creativity. Here's another.

The *BACON* program, put together by Herbert Simon and his group at Carnegie-Mellon University, is another interactive program. *BACON*'s mission is to detect lawful structure in masses of observational data, and extract that lawfulness in the form of mathematical equations that fit the data. In some sense, the idea behind *BACON* is to replace the judgment of a human scientist by a mechanical procedure for identifying relations that compress vast amounts of measured data into a few simple equations.

The innards of *BACON* consist of what's called a *production system*. This is a set of instructions, all of which have the same form, which may be represented as

$$C \to A$$

where the expression C is called the *condition* of the production and A is termed the *action*. The condition side consists of a number of tests; the action side is a sequence of symbol-manipulating actions. Whenever the conditions are satisfied by the current state of the system, the corresponding actions are taken. One further rule is needed to resolve conflicts when two or more conditions are satisfied simultaneously. The simplest such rule is to scan the conditions sequentially, and just execute the action of the first condition on the list that's satisfied.

To see the way such a production system works, here is a particularly simple one that solves a wide range of linear algebraic equations in a single variable x.

$$
\begin{array}{ll}
 & \text{If } x = \mathcal{N} \to \text{Halt} \\
ax + b = cx + d & \text{If } \mathcal{N}x \text{ on right} \to \text{Subtract } (\mathcal{N}x) \\
(a - c)x + b = d & \text{If } \mathcal{N} \text{ on left} \to \text{Subtract } (\mathcal{N}) \\
x = (d{-}b) \,/\, (a - c) &
\end{array}
$$

In this scheme, the system of production rules is in the right column, where the symbol $\mathcal{N}$ stands for any of the constants a, b, c, or d in the equations on the left. In these productions, the symbol x stands for itself. The left column shows the action of the production rule given on the right. In plain English, we can state the first production rule as "If the equation is in the form, an x followed by = followed by a number, then stop." The second rule says, "If there is a term on the right-hand side of the equation having the form, a number followed by x, subtract that term from both sides of the equation." The other rules are read similarly.

One might assume that when schoolchildren learn to solve such equations in classes on elementary algebra, they are acquiring production rules analogous to those given above. Of course, for more complicated equations it would be necessary to augment the set of rules considerably. But the main point is to ask whether such rules underlie *all* cognitive tasks. In the work of Simon et al., evidence in favor of such a hypothesis is put forward by showing how production systems can be used to find regularities in a situation that routinely confronts scientists. This is the problem of finding a mathematical law that fits a given set of measured numerical data. We might think of such a task as an exercise in fitting a curve through an isolated set of data points—or as a creative effort concerned with discovering new laws to describe and explain empirical observations about the world around us. To illustrate this idea, let me describe how *BACON* uses a production system to "discover" Kepler's Third Law of planetary motion.

In his studies of planetary motion, astronomer Johannes Kepler discovered three laws governing how the planets move about the Sun. His first two laws were announced in the year 1609, while the Third Law was discovered nearly a decade later, in 1618. Kepler himself never numbered these laws or distinguished them in any special way from his other discoveries.

These three laws of planetary motion can be stated as follows:

1. All planets move about the Sun in elliptical orbits, having the Sun as one of the foci.

2. A radius vector joining any planet to the Sun sweeps out equal areas in equal lengths of time.

3. The squares of the periods of revolution of the planets are directly proportional to the cubes of their mean distances from the Sun.

Knowledge of these laws, especially the second (the law of areas), proved crucial to Isaac Newton in 1684–85, when he formulated his famous law of gravitation between the Earth and the Moon and between the Sun and the planets, which he postulated to have validity for all objects anywhere in the universe.

It is only the last of these three laws that involves a precise numerical relationship, the first two being simply qualitative statements about the geometry of planetary orbits. We can mathematically write Kepler's Third Law in the form

$$P = KD^{3/2} \tag{*}$$

where P is a planet's period of revolution about the Sun, D is its average distance from the Sun, and K is a constant characteristic of the particular planet whose orbit is under investigation. At the time of Kepler's work, only the measured data on planetary periods and distances (available from the observations of the Danish astronomer Tycho Brahe) could possibly offer a clue as to the form of relationship between P and D given in (*). So how could Kepler have found it?

Actually, there are several possible routes to the expression (*). For example, one might graph the data points and then notice that the relationship between P and D is curvilinear. Straightening out this curve by regraphing the data on a different scale would then suggest the formula (*). But *BACON* follows a different route to the same result.

BACON's first heuristic is to search for correlations between the observables. This rule turns up the fact that as P increases, D decreases. On the basis of this observation, the program tests whether the ratio between P and D, $P/D = R_1$, is a constant. It isn't; but *BACON* retains the quantity R_1 as a new variable anyway.

Next, *BACON* notices that R_1 varies with D. So it computes the new ratio between them:

$$R_1/D = P/D^2 = R_2$$

Again, R_2 is not constant, but varies inversely with R_1. So *BACON* then tests the *product* of R_1 and R_2, finding that

$$R_1 R_2 = P^2/D^3 = K$$

a constant. Thus, *BACON* has found an invariant function of the observables P and D, namely, the relationship $P^2 = KD^3$—exactly Kepler's Third Law.

It's instructive to analyze the two simple heuristics that the program used to arrive at this result. They are:

• If two quantities vary proportionately (inversely), test their ratio (product) for invariance.

• Retain the ratios (products) so obtained and treat them as new variables. Continue to apply the first rule to new pairs of variables.

Of course, *BACON* doesn't actually *know* anything about planets, orbits, gravitation, energy, or any of the other things that a human astrophysicist would bring to bear upon the problem of teasing out a mathematical relationship between a planet's period of revolution and its distance from the Sun. In fact, Newton showed that the motion of bodies subject to central gravitational force need not always follow the elliptical orbits specified by Kepler's First Law but can take paths defined by other conic curves—parabolas or hyperbolas, depending on the total energy of the body. Thus, an object of sufficient energy—a comet, for instance—can enter the solar system and leave again without returning. *BACON* will never know about this, because it operates only on data from planets, not comets, meteorites, or any of the myriad other bodies moving about out there in interplanetary space. Nevertheless, a significant conclusion to be drawn from this work is that relatively simple curve-fitting methods may be enough to discover important scientific laws—even without guidance from theory.

The *BACON* program is an example of what we earlier termed a *microworld*. It is supplied with a few primitive concepts in a restricted domain, along with the goal of creating new concepts in that domain. The program is then given some conjectures relating to those concepts, as well as some criteria for evaluating whether a concept is "interesting" or "uninteresting." One such rule, for example, would label a concept "interesting" if it were independent of all previous concepts known to the program. Finally, a number of heuristic rules are provided to aid in searching for new concepts.

Other interesting efforts in this direction are the programs *Automated Mathematician* (*AM*) and its successor, *Eurisko*. Developed at Stanford University by Douglas Lenat in the early 1980s, both of these programs attempt to generate new discoveries in the domain of mathematics. Their output was impressive. In a few hours of computing, *AM* reinvented the integers and the operations of arithmetic (addition, subtraction, multiplication, and division), the prime numbers, as well as a number of "interesting" mathematical conjectures, such as Goldbach's Conjecture to the effect that every even integer greater than 2 is the sum of two prime numbers. But, again, neither *AM* nor *Eurisko* actually knew about mathematics; they had no knowledge of the real world to draw upon in generalizing these results outside the very specific domain of arithmetic, a domain that was essentially taught to them by feeding them a lot of elementary concepts and rules. Encouraged by these results in the restricted domain of mathematics, Lenat took on the vastly more far-reaching task of trying to give a computer common sense.

Cyc

Suppose you are given the two sentences

Time flies like an arrow.

and

Fruit flies like a banana.

Humans have no difficulty knowing that the first sentence can be understood only as a simile, while the second is a statement of fact. Anyone with a commonsense knowledge of the world can immediately parse these sentences correctly. But as in the example given earlier of the laughable computer translation of the phrase "The spirit is strong, but the flesh is weak" into "The vodka was good, but the meat was rotten," common sense is not one of the computer's strong points. Lenat set out to change this deplorable state of affairs in an effort called the Large Common Sense Knowledge Base project, termed, for brevity's sake, *Cyc.*

The *Cyc* project had its inception when Alan Kay, one of the legendary figures in the computing business, was planning a computerized encyclopedia while working at the Atari research center in the 1970s. When Kay asked Lenat to add something to the encyclopedia, just to jazz things up, Lenat suggested "automating the white space." By this, he meant computerizing the commonsense knowledge that we tacitly assume in making our way through everyday life—such as paging through an encyclopedia. This includes information like the fact that water does not flow uphill or that a person cannot be in two places at once.

When Atari hit financial difficulties in the early 1980s, Lenat moved his idea to the Microelectronics and Computer Technology Corporation (MCC) in Austin, Texas. There MCC researchers regarded *Cyc* as a decade-long project, one that would be several times larger than existing expert systems. Lenat recognized immediately that there would be no shortcut in the process of making a machine intelligent, and that they would need to construct a robust architecture for *Cyc* that could adapt easily to changes.

The MCC group initially decided to represent knowledge in frames, which are mini databases that collect related pieces of information together in a structured way. But it soon became clear that this way of structuring knowledge was too limited to express the type of taken-for-granted knowledge that *Cyc* was trying to assemble. So Lenat and others developed an entirely new language, *CycL,* a combination consisting of a form of logic called the *first-order predicate calculus,* frames, and a few other techniques. This language enables commonsense knowledge to be gathered as a collection of assertions that are true by default rather than by definition. So, for example, instead of holding that "all birds fly," *CycL* holds that birds are creatures that "usually fly."

A vast army of "data enterers" around the world feed the *Cyc* database with information. But it's not the kind of information you would normally see in an encyclopedia or almanac. For instance, *Cyc* knows nothing about Bill Gates being president of Microsoft or that Moscow is the capital of Russia. But it does know what a capital city is and what it means to be a president of a corporation. Introducing actual facts, such as these, into the *Cyc* database will occur at a later stage of the project.

And what will signal the end of the *Cyc* project? Well, it will come when *Cyc* begins to read and learn for itself. This process has actually already begun, since the program now gets almost half its information by scanning texts and extracting new knowledge from them to incorporate into its database. Nowadays, researchers mostly need only explain to *Cyc* sentences such as those above involving "time" and "fruit flies."

The next stage of the *Cyc* project is to add applications. Applications currently envisioned range from automated brokering in financial exchanges to "smart" spreadsheets that would check entries in the rows and columns of the spreadsheet for violations of common sense. Such data might involve entries in which a person had listed himself as a contact in case of an emergency. Once such applications are in hand, it would be safe to declare *Cyc* a success.

Cyc is not without its detractors, however. Foremost among them is Hubert Dreyfus, who describes *Cyc* as the last bastion of top-down strong AI. According to Dreyfus, *Cyc* will fail to produce humanlike intelligence for the same reason that all other top-down efforts have failed: the erroneous belief that the mind is simply a symbol manipulator. As Dreyfus notes, when we read the sentence "Mary saw a dog in the window and wanted it," we know that "it" refers to the dog and not to the window. And we know this not by consulting a database of common facts, but because we understand the emotions Mary feels when she sees the dog. This line of argument says that humans bring their situated experience to bear on understanding language and in reasoning about the world.

Lenat pooh-poohs this objection, likening it to the religious view that intelligence requires a soul. He then goes on to say that if situated elements of the sort underlying Dreyfus's argument add power to *Cyc,* let's throw them into the database as well. "It's not the vindication of one ideology or another," says Lenat. "We're after power." Ultimately, Lenat sees *Cyc*-type databases as something that will underlie all intelligent computing. "In twenty years you might not dream of buying a computer that does not have common sense underlying every application."

Cyc is probably the grandest attempt yet to teach a machine about the world around it. But it's not the only way. And, in fact, bottom-up researchers have rediscovered one of the first approaches to this problem put forth in the early days of AI in the 1950s. This is simply to try to mimic the *hardware* of the brain in silicon and software. This approach leads to the bottom-up argument for machine intelligence based upon what have come to be called *neural networks.*

Matter Matters

In the early days of AI, the top-down, symbol-processing view of intelligence and the bottom-up, neural network approach competed on a more or less equal footing. But in the late 1960s, Marvin Minsky and Seymour Papert published a mathematical account of the limitations of the *perceptron,* a neural network system that learned to recognize the letters of the alphabet. This critique dealt a "devastating" blow to the neural network researchers, resulting in the drying up of research support, hence students, for this approach to creating machine intelligence. Only some decades later was it properly recognized that the Minsky and Papert "impossibility" result only applied to a very particular type of neural network under severely restricted circumstances, and in no way represented a broad-brush condemnation of the entire bottom-up approach to machine intelligence. And, in fact, with the top-down research running up against major barriers occurring at about the same time that high-power computing capability was becoming widely available, the bottom-up view of AI began to make a resurgence in the mid-1980s.

To restate the matter, the basic credo of the bottom-up school is simply that if you want to mimic the performance of a human brain, then perhaps you ought to think seriously about not only what that organ does, but its physiological structure. Perhaps that structure of the brain is important in cognition, and so if you want to build a software version of the brain that carries out the same sorts of cognitive tasks, you ought to respect the brain's physiology in that effort. So, unlike the top-downers who say that one can just skim cognition off the top of the brain, ignoring its actual physical structure, bottom-up researchers say that the structure is essential in the brain's behavior. In short, matter not only matters for cognition, it is essential.

One researcher who has embraced the bottom-up view of intelligence is AI guru Marvin Minsky of MIT. Minsky advocates the position that the mind is made up of many individual modules, each of which specializes in a different aspect of intelligent behavior. So, for example, there is a module for doing arithmetic, another module for spatial reasoning, and yet a third module for tying your shoelaces. In this "society of mind" approach to AI, intelligence emerges from the interactions of conflicting, competing parts in a fragmented mind. Minsky's model represents a radical departure from the rule-following, central-actor paradigm beloved by the top-down advocates of machine intelligence.

As noted above, an artificial neural network consists of "modules" that have some of the properties of real neurons, foremost among them being many inputs, some of which excite the module and some of which inhibit it. The module then takes a weighted sum of these inputs and produces a single output if the weighted sum exceeds a given threshold. Rather than being built into the hardware, these networks are generally simulated in software on a computer. The net is characterized by the properties of the units that compose it, the way the units are connected together, and the procedures by which we change the strength of those connections. The overall structure of such a net was shown earlier in Figure 5.4.

One of the first things researchers learned about the properties of these networks is that memory could be stored in them in a fashion entirely different from memory storage in a typical digital computer. In a computer, storage of information takes place rather like in a large post office, where bits of information are placed in definite boxes, each having its own individual address. The computer then "remembers" by tracking down the individual bits of information composing the item to be recalled and assembling them, address location by address location, in a completely serial fashion. In a neural network, memory is not localized to specific boxes; rather, it is globally stored over the entire network in the various modules. What is stored, then, is the capacity to produce a particular pattern of output activity in a group of artificial neurons when a particular stimulus pattern is presented at the input layer of neurons. So, a neural network is in essence a pattern-recognition device. Let's take a moment to see how a typical neural network is trained to recognize written characters.

The 64 cells in the 8-by-8 grid shown in Figure 5.7 are used to define a character, like the character "3" shown in the figure. The 64 cells of this grid can serve as the input neurons to our network, with each neuron receiving the signal "1" if the cell has a line in it, or a "0" if the cell does not. The output neurons are chosen to represent the possible characters in the alphabet we want to use. For example, if we consider the usual English alphabet with its 26 upper- and lower-case letters, 10 numerals, and the dozen or so punctuation symbols, we might choose 75 output neurons, each one representing one of these symbols. So, for instance, if the network had learned to recognize the

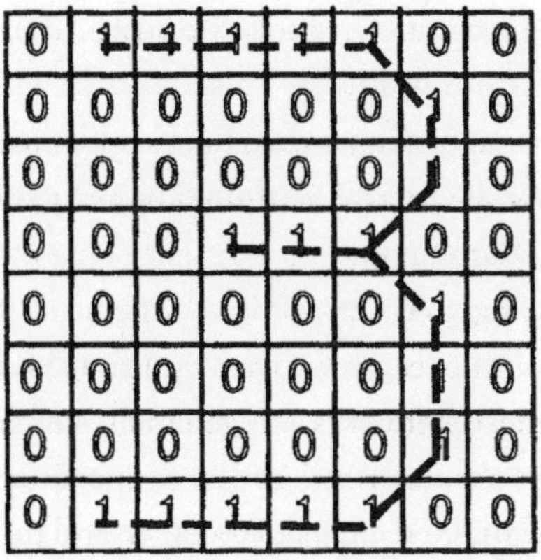

Figure 5.7. Grid for representing alphabetic characters

character "3," the output neuron representing this symbol would fire when "3" is presented at the input. Between the 64 input neurons and the 75 output neurons is a hidden layer of neurons, as shown earlier in Figure 5.4. These neurons mediate between the inputs and outputs, and it is the changing of the weights on the links between the hidden layer of neurons and the input and output neurons that constitutes learning the pattern of the various characters in the alphabet. How is this learning carried out?

It turns out that there are many ways that the weights on the links in the neural net can be adjusted, so as to train the network. But perhaps the most well-studied method for adjusting the weights is by a procedure termed "back propagation." This is a method in which the error the network makes in attempting to recognize a given pattern is fed back through the network altering weights as it goes, so that the same error doesn't happen again. The precise details of the formula by which the weights change are not especially important for us here. The main point, though, is that there are systematic procedures for updating these weights that enable the network to learn to recognize the various characters in the alphabet. The great drawback to this method is that there is no good theory telling us how many neurons one needs in each layer to learn a particular class of patterns, or how long the learning process will take. So, research in neural networks is much more of an art than it is a science. And this is not the only problem with thinking of neural networks as a kind of artificial brain.

Using algorithms like back propagation to train the network is called "supervised" learning, since there is a known goal that the network is trying to reach (for example, recognition of the character "*3*"). Critics note that the real-world brain inside our heads has no such goal(s); it simply learns various patterns without the benefit of a teacher. So a big challenge for neural network researchers is to create algorithms enabling the net to extract meaningful features from a set of input patterns.

The neural nets fail to correspond to everyday experience in other ways as well. For example, we all know that we can unconsciously modulate the ease with which we recognize certain patterns by focusing our attention on them. But how can we adjust the weights in a neural network in such a way as to mimic this commonplace experience of "paying attention"? And what about the very common situation when our cerebral cortex tells us it simply cannot make sense of an input that it has perceived—that no single built-in pattern seems a better match to the input stimulus than any other pattern? Connectionist researchers like David Rumelhart of Stanford University, Geoffrey Hinton of Carnegie-Mellon, and Igor Aleksandr of the University of London are attacking these and other questions, almost daily offering new algorithms and procedures by which to bring the networks more into congruence with the way we observe the brain to function. But the brain has had a head start of a billion years or more, so perhaps it's not surprising that the computer neural networks and the networks inside our heads are still light-years apart. What is surprising, though, is how much of that gap the artificial networks have already closed in the decade or two that they have occupied center stage on the AI researcher's agenda.

We have looked at the current state-of-play in both the top-down and bottom-up camps' research programs for creating strong AI. Now let's turn to recent arguments from the Defense against the possibility that a computing machine will ever think like a person.

THE APPEAL:
ARGUMENTS FOR THE DEFENSE

Did You Say Chinese?

At a conference in Montreal in 1995, Pat Hayes of the University of Illinois and Kenneth Ford of the University of West Florida launched a broadside against the Turing test for recognizing machine intelligence. According to Hayes and Ford, the test is "harmful" to AI, "damaging its public reputation and its own intellectual coherence." They go on to say, "We must explicitly reject the Turing test in order to find a more mature description of our goals— it is time to move it from the textbooks to the history books." What is the basis for such a strong statement against what seems to be the perfectly sensible idea that if a machine behaves as if it's intelligent, then it is intelligent?

One of the problems with the Turing test, say researchers, is that a program playing Turing's Imitation Game is trying not so much to display comprehension of language as to wield it with an astonishing degree of sophistication. To pass a full-blown Turing test, a program would require much more than ordinary conversational capabilities. It would have to be an expert on making a good impression, it would have to be clever, lie, cheat, and dissemble. So, what such a program would amount to would be not a simple artificial intelligence but more like an artificial con artist.

Critics say that having such a well-defined aim as passing a Turing test is the sign of an immature discipline. What is the Turing test for civil engineering or physics? they ask. As Pat Hayes put it, "Five hundred years ago alchemists sought to pass their own Turing test by searching for the philosopher's stone, a magical stone or substance that could transmute base metal into gold. But they only became chemists after they stopped looking for that stone. As a call to arms, Turing's paper was a magical success. But that was 1950. It's time to put alchemy behind us."

Recalling the Loebner test discussed earlier, what the successful programs tend to rely on in formulating their replies to interrogators is programming tricks of the type considered above. Ironically, one of the competing programs was declared "not human" by a female Shakespearean scholar because its answers on literature were too detailed. It seems as if the ability to produce paragraphs of well-written English is now considered to be an inhuman quality!

All the programs entered in the Loebner competition are what computer scientists call expert systems. But even the best of these systems, like *MYCIN,* a program that answers questions about bacterial blood infections, know nothing beyond their (very) limited area of expertise. They are completely incapable of negotiating their way through life.

But many AI workers feel that Hayes and Ford are wrong to blame the

Turing test for the woes of their subject. John McCarthy, who has been in the
AI game since the field's inception at the famous Dartmouth Conference in
the early 1950s, believes that not even Turing himself regarded the Imitation
Game as a definition. He notes that Turing's other writings on machine intelligence don't even mention it. But others who want to stick to the test dismiss
the objections of Hayes and Ford, arguing that the major problem is taking
Turing too literally. Robert Epstein, of the Cambridge Center for Behavioral
Studies in California, says that Turing "never pretended to give the details of a
practical test."

We noted in an earlier section that one of AI's bitterest critics is Berkeley
philosopher John Searle, inventor of the infamous Chinese Room argument
against the adequacy of the Turing test to distinguish mechanical thought. In
terms of this experiment, even if a program did pass the Turing test it would
still not be displaying humanlike intelligence. In Searle's words:

> Formal symbol manipulations by themselves don't have any intentional-
> ity, they are quite meaningless. They have only a syntax but no seman-
> tics. Such intentionality as computers appear to have is solely in the
> minds of those who program them and those who use them, those who
> send in the input and those who interpret the output.

Objections to the Chinese Room have been hot and heavy ever since
Searle first advanced the argument in 1980. Probably the most common
counter is the so-called Systems Reply, which states that while the person
inside the Chinese Room doesn't understand Chinese, the *entire* system—
consisting of the room, the person inside, and the dictionary—does understand the language.

One of the more interesting counterattacks comes from computer scientist Elhanan Motzkin, who asks, How do we know the person inside the
closed room is thinking in, say, English and not in Chinese? The obvious
answer is that Searle has assumed this at the outset. But this assumption has
momentous consequences, since assuming that our man does not understand
Chinese translates into the assumption that the computer does not understand the meaning of its program (which we see from the outside to be simply
the rules specifying the dictionary by which the person inside the room
matches one Chinese ideogram with another). But this was what Searle's

argument was supposed to prove! The whole point of Turing's Imitation Game was that we should assume no such thing—that comprehension should be judged from the outside, by objective tests. So what Searle has done is assume from the very beginning that the machine is not intelligent, which, of course, makes it a whole lot easier to prove that it isn't and impossible to prove that it is.

The other possibility is simply to ask the person in the room if he or she understands Chinese. If asked in English, the person would say that he or she doesn't. And if asked in Chinese, the person would say that he or she does. (Recall: The point of this game is to simulate a native speaker of Chinese, and such a person would certainly claim to understand Chinese.) So how are we to decide which is the person's true language? To put it simply, we can't!

Motzkin concludes then that Searle's argument falls apart, because we cannot know that our person inside the room doesn't *really* think in Chinese. In countering this line of reasoning, Searle says let the person inside the room be him. Then, he goes on to note that it is a plain and simple fact of life that he, John Searle, does not understand Chinese. Moreover, just having a bunch of rules for manipulating Chinese characters would not teach him the meaning of any of the ideograms. Searle then says that Motzkin thinks that the crux of the question is,"How do we know our man is thinking in English and not in Chinese?" But, Searle argues, "How do we know?" is *not* the crux of the matter. Rather, the issue at hand is "Under what conditions does a system in fact have understanding, regardless of how observers outside the system can tell, or whether they can do so at all?" In the end, Searle falls back on the irreducible core of the Chinese Room argument, namely, that just having a formal syntax is not sufficient for understanding; syntax is not semantics. According to Searle, Motzkin ignores this logical point entirely.

A fitting ending to the discussion over what the Turing test is and is not is the fact that the Loebner prize has been canceled. The prize committee is now preparing to launch a new competition designed "in the spirit of Turing's original proposal." *Sic transit gloria mundi.*

Gödel's New Clothes

Roger Penrose is an Oxford don who has received just about every honor that a mathematical physicist can aspire to, among them the prestigious Wolf Prize (with Stephen Hawking) and a knighthood from Her Majesty, Queen Elizabeth II (but, strangely, not a Nobel Prize—yet!). Penrose grew up in about as intellectual a family as one can imagine. His father, Lionel Penrose, was a famed geneticist, his mother was a physician, and one of his brothers, Oliver, was British chess champion ten times. In fact, someone told me a story (possibly apocryphal) to the effect that mental chess was a popular pastime over coffee and the Sunday papers in the Penrose household. Apparently, moves would be called out across the room from one Penrose to another during interludes between reading sections of *The Sunday Times.* And I can't think of a single scientist who has so many discoveries named after him—Penrose tiles, the Penrose staircase, the Penrose matrix inverse, and a few others relating to quantum theory and cosmology. So with this sort of background, when Roger Penrose speaks, people listen—at least when he speaks on matters of science and the intellect.

In 1989, Penrose spoke on the matter of thinking machines in his book *The Emperor's New Mind.* This volume advances the view that strong AI is impossible. When someone with the credentials of Penrose states a view like this that so many people in both the scientific and general public desperately want to hear, it should be no surprise to discover that the book remained on *The New York Times*'s best-seller list for several months. What is a surprise, though, is the argument that Penrose puts forward to support this comforting conclusion. He bases his entire case on an appeal to Gödel's Incompleteness Theorem, the very same argument that Penrose's Oxford colleague John Lucas published in 1961. It makes one wonder if there's something about the air in Oxford that periodically brings out these Gödelian-based statements against AI. We have sketched earlier the nature of the appeal to Gödel as an anti-AI argument, so I won't repeat it here. Rather, let's look at some of the reasons why many claim that Gödel's Theorem is simply irrelevant to the issue of mechanical thought.

Gödel's result says that if you have a system of axioms and rules of logical inference that you employ to prove mathematical propositions, if the system is consistent there must exist some proposition that can be neither proved nor disproved using that logical framework. And, in fact, Gödel showed that such an "undecidable" proposition could be constructed so as to be seen to be true by looking at it from outside the logical proof system. Both Lucas and Penrose jumped on the existence of such a true—but unprovable—statement to conclude that since every computing machine is a logical system, there must be something that we humans can see to be true, but which the machine cannot prove. Ergo, machines will never be able to think like humans. QED.

Well, almost everything is wrong with this argument. And, in fact, Alan Turing, himself, anticipated it—and demolished it—in his original paper on computing machines and intelligence in 1950. But there are even stronger counterarguments than Turing's. Let me list three of them.

• *Finiteness:* All real-world computing machines are ultimately subject to the laws of physics when carrying out their appointed computational rounds. Among other things, this means that they are *finite* devices. In particular, there are finite slices of time within which a single computational operation can be executed, and the machines all have a finite number of locations where information can be stored. By way of contrast, the equivalence between the *theoretical* model of computation proposed by Turing, which forms the basis for what people like Penrose and Lucas regard as a "computing machine," and Gödel's Incompleteness Theorem relies on the existence of an infinite set of entities. In the case of the Turing machine, this infinity shows up in the assumption that Turing's "paper computer" has an infinitely large memory capacity; in Gödel's Theorem the infinity is the countable infinity associated with the integers, the so-called "counting" numbers. Delving a bit deeper into the arguments that make Gödel's magic work, one quickly discovers that the theorem can *only* work when there is some kind of infinity on the scene; for finite systems, there is—and cannot be—a Gödel type of incompleteness.

Since any computing machine built of matter and powered by energy is necessarily a finite object, the whole notion of Gödelian incompleteness is simply a red herring. The theorem does not apply to such devices, and, hence, is simply irrelevant to the issue of whether or not a computing machine can think like a human. Case closed! But to hammer just a couple more nails into the coffin of incompleteness and AI, let's look at two more objections to Penrose's argument.

• *Metamathematics:* The true sentence that Gödel asserts must exist but be unprovable is something whose proof lies *outside* whatever fixed, logical framework we are using to prove statements about numbers. The heart of Penrose's argument is that the machine must always do its mathematics within such a fixed logical system. One mistaken notion underlying the belief that a computer program can't do mathematics like a human is just this assumption. The following dialogue (courtesy of John McCarthy) between Penrose and a mathematics computer program illustrates the point:

PENROSE: Tell me the logical system you use, and I'll tell you a true sentence you can't prove.

PROGRAM: You tell me what system you use, and I'll tell you a true sentence you can't prove.

PENROSE: I don't use a fixed logical system.

PROGRAM: I can use any system you like, although mostly I use a system based on a variant of ZF and descended from 1980s work of David McAllester. Would you like me to print you a manual? Your proposal is like a contest to see who can name the largest number with me going first. Actually, I am prepared to accept any extension of arithmetic by the addition of self-confidence principles of the Turing-Feferman type iterated to constructive transfinite ordinals.

PENROSE: But the constructive ordinals, like 1st, 2nd and so on, aren't recursively enumerable.

PROGRAM: So what? You supply the extension and whatever confidence I have in the ordinal notation I'll grant to the theory. If you supply the confidence, I'll use the theory, and you can apply your confidence to the result.

The point of this little interchange is not that the computer is inconsistent, per se, but that it can shift from one formal logical system to another during the course of its mathematical work. In short, it can look at mathematical systems from the outside, thereby having the ability to look at Gödel's Theorem as just an ordinary theorem of a higher, metamathematical theory.

• *Consistency and evolution:* No theorem in mathematics is universally true without any assumptions; Gödel's results are no exception. The principal assumption underlying his incompleteness result is that the logical system one is using to prove theorems is consistent. That is, both a statement and its negation cannot both be theorems of the system. In replying to the earlier criticism of strong AI by Lucas, C. H. Whitley noted that Lucas assumed that the human mind is consistent. In fact, this is far from obvious, as the following paradox constructed by Whitley shows.

Consider the sentence "Lucas cannot consistently assert this sentence." Lucas cannot assert the truth of this sentence even though he can clearly see that it's true. Why? Because if Lucas could assert it, then that fact would undermine his assumed consistency. Thus, either there is something that Lucas can see to be true but can't assert, or he is inconsistent. Consequently, Whitley claims, Lucas holds too high a regard for humans, since even if there is an unprovable statement that a specific machine cannot assert, humans can't always do it, either.

Other arguments countering Penrose and Lucas claim that they err in their application of Gödel's results. For instance, the Incompleteness Theorem shows that a machine M cannot prove the true, but unprovable, Gödel sentence of M from *its* axioms and according to *its* rules of inference. But neither can the human mind prove its Gödel sentences from *its* axioms and rules of inference. Furthermore, Penrose doesn't show that he can find a flaw in any machine, but only in any machine that the mechanist can construct.

It's well to conclude this discussion of anti-AI arguments based on Gödel's results by hearing Gödel's own view on the matter. Unfortunately, Gödel was rather reclusive and secretive, especially in his later years, and his only published statement on the topic comes from a lecture delivered to the American Mathematical Society in 1951:

The human mind is incapable of formulating (or mechanizing) all its mathematical intuitions, i.e., if it has succeeded in formulating some of them, this very fact yields new intuitive knowledge, e.g., the consistency of this formalism. This fact may be called the "incompletability" of mathematics. On the other hand, on the basis of what has been proved so far, it remains possible that there may exist (and even be empirically discoverable) a theorem-proving machine which in fact *is* equivalent to mathematical intuition, but cannot be *proved* to be so, nor even be proved to yield only *correct* theorems of finitary number theory.

Thus Gödel leaves open the possibility of the existence of a theorem-proving machine, and even concedes that it may be possible to discover such a machine by empirical investigation. However, he then throws a wet blanket on the whole business by saying that if we ever find such a machine, it will be beyond our powers to prove that it constitutes a Universal Truth Machine!

THE APPEAL:
SUMMARY ARGUMENTS

Both sides of the strong-AI debate have made telling arguments supporting their positions. On the Prosecution's side favoring the possibility of thinking machines, the top-down and bottom-up camps have pushed forward their respective research programs with great vigor. The Defense witnesses have also sharpened their claims against the possibility of our ever attaining the goal of constructing a machine that could think like a human. Here is a tele-graphic, tabular summary of their respective positions.

YES, COMPUTERS CAN THINK!

PROMOTER	RESEARCH PROGRAM
Top-down School	
Simon and Newell	*BACON, Racter*
Lenat	*Cyc*
Bottom-up School	
Rumelhart, Aleksandr	neural networks
Minsky	"society of mind"

Table 5.3. Summary arguments for the Prosecution

NO, COMPUTERS CANNOT THINK!

PROMOTER	ARGUMENT
Searle	Chinese Room
Dreyfus	phenomenology
Lucas, Penrose	Gödel's theorem

Table 5.4. Summary arguments for the Defense

THE APPEAL:
THE DECISION

Comparing Tables 5.1 and 5.2 with Tables 5.3 and 5.4, it is evident that there have been no dramatically new ideas in the AI debate over the past decade; there are no new contenders in either camp. Consequently, the issue of whether the appeal of the 1989 decision in favor of the Prosecution's bottom-up arguments should be granted rests on whether any significant new evidence has come to light during the past decade in the previous arguments for either the Prosecution or the Defense. Let's consider the Prosecution first.

If performance constitutes progress, then the top-down advocates of strong AI have a case to answer. We now have a program that can beat the world champion at chess; we have programs like *BACON* that can discover laws of physics from observed data; we have programs like *Cyc* that hold great promise for being able to interact with the world on the basis of a general, not restricted, knowledge base. Of course, opening up the machine and looking in to see how this magic works, one observes nothing that's even remotely close to how we think humans perform the same tasks. So these impressive accomplishments are more in the category of *weak*, rather than strong, AI. Nevertheless, significant progress has been made in bridging the performance gap between the machines and the humans by the top-down workers.

Even more dramatic progress has been made by the bottom-up investigators, as evidenced by the myriad ways that neural networks have entered into both the academic and commercial marketplaces. In almost every area having to do with making money—stock markets, sports betting, customer credit evaluation—commerically developed neural networks are employed on a routine basis. The same holds for other areas such as medical diagnosis, optimization of chemical plants, handwriting recognition, and identification of signals in the presence of noise. Since businessmen don't generally pay for things that don't work, this almost universal use of neural networks is the strongest possible testimony to their utility in providing humanlike behavior in a plethora of problem-solving domains. While there is still a great distance between the pattern-recognition capabilities of computers and actual human intelligence, the eye-opening progress being made with neural nets over the past decade strongly suggests that this bottom-up attack on creating a mechanical mind is still on the right track.

Turning to the Defense, the situation is not nearly so rosy. Certainly, the biggest splash for the Defense has been the appearance of Penrose's Gödel-based argument against strong AI. While his book *The Emperor's New Mind* and its successor, *Shadows of the Mind,* contained a cornucopia of fascinating facts about mathematics, quantum theory, consciousness, neurophysiology, and a whole lot more, the one thing that it was not was a convincing argument against AI. The centerpiece of Penrose's whole case, the appeal to Gödel's Theorem, is a tired, outmoded line of attack that was dismissed by Turing, himself, in 1950. The resurrection of the idea by Lucas in the early 1960s was even more convincingly demolished, so it really makes one wonder how such a brilliant mind as Penrose's can believe that it holds any merit whatsoever. I suspect that if this same argument had been put forward by anyone of lesser visibility and accomplishment than Roger Penrose, the book would have sunk like a stone on its day of publication. But, Penrose or not, the argument is completely irrelevant to the AI question.

Turning to the notorious Chinese Room, John Searle has now spent nearly twenty years explaining and defending his antibehavioristic argument against the adequacy of the Turing test against all comers—but to no avail. In every one of his salvos, Searle marvels at the fact that anyone with even half his or her brain working can fail to see how crystal-clear is his argument against the possibility of a computing machine ever thinking like a human. One can only wonder how there can be so many intelligent commentators who have failed to be moved by Searle's line of reasoning. Can it possibly be that the Chinese Room gedankenexperiment is just as irrelevant to the issue of thinking machines as is Gödel's Theorem? I think so.

So, on balance, we come to the conclusion that nothing of eyebrow-raising substance has really changed in the AI debate since the mid-1980s. Of course, the pro-AI workers have made steady progress on their research programs. But the anti-AI group is still hawking the same old tired goods. No court in the land would grant a new trial on the basis of such feeble nonevidence.

THE DECISION:
APPEAL DENIED

SIX

Who Goes There?

Claim: There Exist Intelligent Beings in Our

Galaxy with Whom We Can Communicate

BACKGROUND

Are We Alone?

Probably the most highly visible scientist of the past few decades was Cornell University astronomer Carl Sagan. Before his death a couple of years ago, Sagan was a tireless campaigner for the view that the universe must be just teeming with intelligent life-forms that we could be in meaningful communication with. From his immensely popular PBS television series, *Cosmos,* to his Pulitzer

Prize–winning book, *The Dragons of Eden,* Sagan used his charm, intelligence, and name-brand recognition to push Congress to fund various searches for extraterrestrial intelligences (SETI). Sad to say, even Sagan's charisma was not enough, and government support for SETI eventually dried up. But the allure of the alien remained alive in enough people that private support was found to continue the search. But what, exactly, is the basis of Sagan's and others' view that there really is intelligent life "out there"? And if their case is so strong, why did the government withdraw its already minuscule level of support? After all, one would certainly think that contact with an intelligent extraterrestrial being would be a world-shattering event. So what are the arguments—pro and con—for the existence of such beings? Let's first look at Sagan's case.

From Copernicus to Drake

Over five hundred years ago, the Polish astronomer Nicolaus Copernicus pushed humans off center stage in the universe when he showed that the Earth is not even the center of our solar system, let alone the universe. This dramatic discovery, that the Earth revolves about the Sun rather than the other way around, serves as the jumping-off point for the argument that intelligent extraterrestrials (ETIs) must exist. The line of reasoning is very simple. Since there is nothing special about our Sun, our planet, or our physicochemical structure, there must exist many such suns, planets, and carbon-based beings elsewhere in the galaxy. And since the laws of chemistry, physics, and biological evolution can be expected to apply equally wherever you are in the universe, it stands to reason that there should exist beings not very different from us in many corners of the galaxy. In other words, we are very mediocre in every possible respect; therefore, there should be lots of intelligent humanoid-type organisms in the Milky Way galaxy. Or so goes the argument advanced by Sagan et al., often termed the Principle of Mediocrity.

In 1960, radio astronomer Frank Drake gave mathematical meaning to the Copernican Principle, when he constructed an equation for estimating the number of advanced communicating civilizations in our Milky Way galaxy. Drake considered the various conditions that would all have to be satisfied for a communicating civilization to exist. They would include the right habitat, the emergence of some kind of life in this environment, the likelihood that that life would be intelligent, and so forth. He then put all these factors together into an equation that would estimate how many such civilizations there are at a given time. Here's a slightly more detailed look at the components making up the Drake equation.

R^* = the rate at which stars are formed in our galaxy per year

f_p = the fraction of stars, once formed, that will have a planetary system

n_e = the number of planets that will have an environment suitable for life

f_l = the probability that life will develop on a suitable planet

f_i = the probability that life will evolve to an intelligent state

f_c = the probability that intelligent life will develop a culture capable of communication over interstellar distances

L = the time (in years) that such a culture will spend actually trying to communicate

Under the very dubious (but simplifying) assumption that each of the foregoing factors is independent of the others, an estimate for N, the number of advanced communicating civilizations in our galaxy, can then be made by just multiplying each of the factors together. This yields the celebrated Drake equation,

$$N = \underbrace{R^* \times f_p \times n_e}_{physical} \times \underbrace{f_l \times f_i}_{biological} \times \underbrace{f_c}_{cultural} \times L.$$

To use this equation in any meaningful sense means that we have to make numerical "guesstimates" for each of the seven terms of the equation. This is where the Copernican assumption comes into play, since there is virtually no scientific evidence upon which to base a rational estimate for *any* of these terms other than the first two. So most investigators fall back onto the argument that there's nothing special about our place in the grand scheme of things, in order to make estimates that are unabashedly anthropomorphic in character. Table 6.1 gives the result of a few such exercises, showing the vast range of possibilities. In the table, the symbols H, M, and L represent high, medium, and low guesstimates of the various quantities.

	SHKLOVSKII AND SAGAN (1966)			HART (1980)			ROOD AND TREFIL (1982)		
TERM	H	M	L	H	M	L	H	M	L
R^*	$-^*$	10	–	50	20	10	0.15	0.05	0.005
f_p	–	1	–	0.5	0.2	0.025	0.30	0.10	nil
n_e	–	1	–	1	0.1	0.001	0.20	0.05	nil
f_l	–	1	–	1	0.1	10^{-20}	0.50	0.01	nil
f_i	–	0.1	–	1	0.5	0.1	1	0.50	nil
f_c	–	0.1	–	1	0.5	0.1	1	0.25	nil
L	$>10^8$	10^7	100	10^6	10^4	100	10^6	10^4	100
N	$>10^8$	10^6	100	25×10^6	100	nil	4500	$\sim 10^{-3}$	nil

*No upper or lower estimates given

Table 6.1. Estimates for N using the Drake equation

About the only thing one can conclude from these numbers is that N can be any number you want! Just concoct a plausible scenario for why one of the quantities has a certain value. The end result for N can then be anything between 1 and several tens of millions. Not exactly the kind of precision that the general public looks to the scientific community to provide!

But the Drake equation is not entirely useless. It has the singularly help-ful feature of focusing attention on many factors—astrophysical, biological, cognitive, sociocultural—that propel a civilization to the stage where it can communicate with the stars. If nothing else, the equation set a research agenda for the entire SETI movement, replacing what had previously been mere scholastic squabbling. Now to the loyal opposition. On what pillar(s) does their argument rest for claiming that $N = 1$?

Where Are They?

The legendary Italian physicist Enrico Fermi once asked the question, If the universe is overflowing with intelligent life-forms just like us, where are they? What Fermi had in mind was the simple fact that if ETIs are out there in vast numbers, then at least some of them would be technologically much farther advanced than we. Moreover, their curiosity would have led them to construct devices for exploration of the universe, either manned spacecraft or, far more likely, unmanned—but intelligent—probes. So, asks Fermi, why haven't we seen a single shred of unequivocal evidence from these spacefaring civilizations? Appealing to Occam's razor, by which the simplest explanation of an observed fact is the preferred answer, Fermi's answer is that ETIs just don't exist.

Fermi's answer is what I call the reductio ad absurdum argument for $N = 1$. Here is the logic underlying it.

1. Assume N is greater than 1 and deduce the consequences of this assumption.

2. If the consequences are not observed, then we must conclude that the original assumption is wrong and that $N = 1$, after all.

3. In this case, the consequence that follows from N being much larger than 1 is that we should long ago have seen undisputed evidence of extraterrestrial civilizations. For example, the appearance of a von Neumann probe, an unmanned, instrumented probe that would observe earthly goings-on and signal this activity back to the alien home planet or mothership.

4. But, alleged UFO sightings aside, no such evidence has ever been observed or reported. Hence, N is not greater than 1, after all.

But there is another route to the same conclusion. It is through the Drake equation.

All that's needed for N to be vanishingly small is for just one term of the Drake equation to be very small. In that case, the entire product of all seven terms will be small, and we can then conclude that we are alone. Many investigators, especially biologists, have followed this route. For example, the eminent evolutionary biologists Ernst Mayr and George Gaylord Simpson claim that the likelihood of life-forms emerging at all, and, if they do, of developing intelligence, is microscopically small. Hence, according to the arguments of Mayr and Simpson, the terms f_l and f_i in the Drake equation would both be small. Similarly, astronomer Michael Hart has argued that n_e, the number of habitable planets surrounding a star, is vanishingly small. And in an entirely different direction, the philosopher of science Nicholas Rescher has advanced the claim that the likelihood of our ever being able to communicate with an ETI is essentially zero (witness our feeble efforts at communication with intelligent species on our own planet, like dolphins and whales).

Taken together, the reductio ad absurdum and the "factorization" arguments against the Drake equation create a devastating case in support of the Defense claim that $N = 1$ (ETIs do not exist). Tables 6.2 and 6.3 summarize the state of play for both sides circa 1988.

N>1: ETI EXISTS!

PROMOTER	ARGUMENT
	N *is large*
Sagan, Morrison	Principle of Mediocrity
	N *is small or large*
Dyson	comets or Dyson spheres
Papagiannis	asteroid belt
	N *is moderate*
Drake	travel/colonization too expensive
	Agnostic
Rood	Drake equation
Bracewell	von Neumann probes

Table 6.2. Summary arguments for the Prosecution

$N = 1$: ETI DOES NOT EXIST!

PROMOTER	ARGUMENT
Hart	no colonization; f_e small
Tipler	absence of von Neumann probes
Mayr, Simpson	f_l, f_i, L small
Trefil	no colonization
Carter	Anthropic Principle
Rescher	otherworldly science

Table 6.3. Summary arguments for the Defense

With these arguments from a decade ago under our belts, let's turn to recent developments that might tend to shift the odds. The logical place to start is with the physical environment, since if an ETI like us is to exist "out there," it's going to have to be in an environment not too dissimilar from our own.

A Plethora of Planets

In 1996, Daniel S. Goldin, the administrator of NASA, announced the space agency's Origins program, which is aimed at finding evidence of extraterrestrial life. In his announcement, Goldin held out the hope that one day astronomers would build a telescope "powerful enough to photograph the surface of an earthlike planet orbiting a distant star, with enough resolution to distinguish clouds, continents, oceans." Good luck, Mr. Goldin! Such a vision, appealing as it is to the untutored eye, is pure fantasy. Even if astronomers could somehow isolate a planet from the glare of its parent star, they could not see such detail. Every telescope blurs the fine points of an image because the incoming radiation is a wave that spreads as it enters the telescope. But this does not mean that we cannot discover the existence of planets surrounding stars other than our own Sun. And, in fact, discoveries of this very sort have been among the most exciting developments in SETI over the past decade.

The first detection of a planet around a Sun-like star occurred in 1995.

Michel Mayor and Didier Queloz of the Geneva Observatory in Switzerland found a Jupiter-sized object in an orbit less than one-sixth the radius of Mercury's. That planet, around the star 51 Pegasi, turned out to be the first of a series of "hot Jupiters"—giant planets far closer to their parent star than standard theory predicted they should be. Later, nine more planets were discovered that included objects so massive, in orbits so eccentric, that theorists are hard pressed to imagine how they could form at all. To further muddy the waters, other extrasolar planetary discoveries have the look and feel of our own solar system. So how do astronomers find such objects, when direct observation is impossible?

The answer is simple: indirect deduction. Basically, the method is to sift through hundreds of dark features called absorption lines in a star's spectra. If the gravitational pull of an orbiting companion is making the star wobble, like a slightly unbalanced washing machine, the Doppler shift will cause these lines to creep back and forth. This wobble can then be interpreted to give the smallest possible mass for the companion object—0.47 times the mass of Jupiter for the companion to 51 Pegasi, provided we are viewing the companion head-on. But if we happen to be seeing the orbit nearly face-on, the object's mass would have to be much larger, perhaps as large as a star's, to produce the same wobble.

Some investigators have questioned whether these companion objects are really planets. For instance, the companion to 51 Pegasi is so close to the parent star and so large that some feel it is really the junior partner in a binary star system that never quite got ignited as a full-fledged star. Such massive, Jupiter-sized objects that are within a hairsbreadth of being stars themselves are called brown dwarfs, in recognition of the class of small, bright real stars, the white dwarfs.

But even before the discovery of planets around Sun-like stars in 1995, in 1992 Aleksander Wolszczan of Penn State University found three Earth-size objects in orbit around a pulsar, a fast-rotating neutron star. By causing the pulsar to wobble, the planets create periodic changes in the otherwise clocklike regularity of the pulsar's radio outbursts. The precision of Wolszczan's method of measurement is so great that it can discern the gravitational "kick" the planets give to each other as they pass in their orbits. That kick is the real key to identifying these objects as planets.

So the evidence is now overwhelming that planetary systems form around stars other than our own Sun, which is very big news for the SETI community, since prior to these discoveries in the early 1990s the existence of planetary systems was really only a conjecture. It was an axiom of faith that planets would be needed in order for ETIs to have the right kind of environment to develop. So the estimates n_e in the Drake equation (see Table 6.1) are all simply guesses; in the next few years, we will be able to give numbers based on factual observations.

Just like Home

Not every planet is a good candidate for the habitat of ETIs. It must be a planet having the "right stuff." A hot, massive planet like Jupiter could harbor some kinds of life-forms, but it is unlikely such life would be able to communicate and share information with us. For that, we need Earthlike planets. So the real question for SETI workers is, How many planetary systems could form having Earthlike planets as part of their retinue?

A number of researchers have developed sophisticated computational tools for simulating how the Sun's collection of planets formed by accumulation of objects of sizes ranging from a tenth of a kilometer to ten kilometers across. Such objects are often termed planetesimals. Conventional wisdom says that these objects first form into "embryos" about the size of the Moon, then coalesce into planets like Mars, Venus, Earth, and Mercury. The computational experiments vary factors like the surface density of the disk or the mass of the central star, in order to see what types of planetary systems typically tend to form. A typical sampling of such experiments performed by H. Levison and his colleagues in 1998 is shown in Figure 6.1. The horizontal axis shows the distance of planets from the star, measured in astronomical units (1 AU = the average distance between the Earth and the Sun); the markings below the planets show the range of distances covered by each planet in its orbit, while the numbers above the planets indicate their masses (relative to the Earth's mass).

A concept central to all discussions of SETI is the "habitable zone" (HZ). This is a region of space, centered on the central star of the solar system, whose inner and outer boundaries are the distances at which water would boil or freeze, respectively. The underlying premise is that life is water

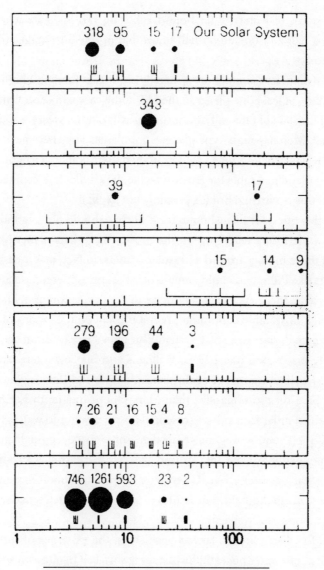

Figure 6.1. Some simulated solar systems

based, so surface conditions on planets within the HZ should allow simple organisms such as bacteria to survive. Of course, the HZ may still be too extreme to be compatible with the presence of more advanced life-forms. More recently, the HZ has been redefined to be the locations where a run-away greenhouse effect would occur (inner boundary) and where carbon dioxide clouds would serve to decrease a planet's temperature below the freezing point of water (outer boundary). The HZ for stars whose size is between a half and one-and-a-half times that of our Sun ranges from about 0.2 AU to 2.8 AU.

The planetary simulations generally show that there is a high probability of finding at least one planet in the HZ in the case of a central star like our Sun. But for stars of one-half or one-and-a-half solar masses, these are usually only small, Mercury-mass-type planets, not planets anywhere near the size of the Earth. Nevertheless, for stars like our own there is at least a fighting chance of finding planets suitable for Earth-type life. It is difficult to estimate the likelihood of this occurring. But it's certainly not negligible.

In this quest for likely homes for ETIs, we should not overlook the possibility of life-forms emerging on the moons of a giant planet like Jupiter. Some of these moons around extrasolar planets, in fact, may well be as large as planets like Pluto and could contain oceans, atmospheres, and all the other paraphernalia theorists feel is conducive to the development of life. Unfortunately, such hospitable conditions are not present on the moons of any of the giant planets in our own solar system. But who's to say about the moons of the newly discovered planets? D. Williams and his coworkers have considered the possible moons of such planets, and found that any moon would need to be a bit more massive than Mars (about one-tenth Earth mass) to retain a substantial atmosphere (essential for surface liquid water) for billions of years. Such a moon might also need an appreciable magnetic field to protect its atmosphere from loss caused by charged particles striking it from either its star or outer space. Ganymede, the largest moon in our solar system, has a magnetic field, but is only 3 percent of Earth's mass. But more massive moons may well exist.

All the discussion so far has been about the existence of surface water. But what about water beneath the planetary surface? In our own solar system, circumstantial evidence for a liquid subsurface ocean on Jupiter's moon, Europa, is growing, and there could be regions of Europa where conditions lie within the range of adaptation of Antarctic terrestrial organisms. But Europa lies well beyond the HZ, has only about a tenth of Mars's mass, and has almost no atmosphere.

What this all adds up to is that we really don't know very much about the conditions necessary for life to get going on another planet. So when we don't know, we tend to fall back upon what we do know, namely, what has happened here on Earth. And this is a very chauvinistic view of the conditions needed for ETIs to emerge. But let's press on nonetheless.

Suppose ETIs do exist on one or another of these planetary systems. It's not unreasonable to suppose that the physical form of such an entity would be greatly influenced by the physical environment. For instance, if the being were one of the cheela from Robert L. Forward's science fiction novel *The Dragon's Egg*, and had evolved on the surface of a neutron star, its physical form would reflect the gigantic gravitational pull and magnetic field strength on the surface of such a star. This form—the cheela had a flattened, pancake-type body—could then be presumed to affect the behavioral pattern of the being. But let's explore this notion further.

What Would Happen If the Tape Were Run Twice?

The famed geneticist Conrad Waddington was convinced that the highest form of life on any Earthlike planet would closely resemble . . . Waddington. Unlikely! Evolution is a contingent affair, and if the evolutionary tape were to be run again on this planet, the land vertebrates would be very unlikely to reappear, and even if something like them did, their anatomies would almost surely have many important differences.

On any other planet, even a second Earth on the other side of the galaxy, the chances that our same genetic system would arise, with the same oddities selected, leading to the same combinations of genes, seem slight. Finding another planet with Earth-style dinosaurs or people is about as likely as finding a remote desert island in the Pacific where the natives speak perfect Afrikaans or rhyming cockney slang.

So if we can't have human beings or dinosaurs on a distant planet, what can we have? After all, aliens cannot be invented out of whole cloth. Although every detail must be different, there are patterns of general problems, and common solutions to those problems, that would apply to life anywhere else in the universe. For example, aerial flight was developed by the ancestors of birds, insects, bats, and certain types of fishes. Another example is photosynthesis, which was invented independently by many different bacteria-type organisms. Yet another evolutionary solution is symbiosis, in which several disparate organisms combine to make a much more

complex reproductive system, as in termites and in cows, which have sym-
bionts in their guts.

It is reasonable to suppose that these universal solutions will be found
on pretty much all other planets with life. But there are also many important
"accidental" contingent inventions, which Waddington and all other land vertebrates have inherited from the fish that came out of the primordial seas many
aeons ago. These parochial oddities are specific to one evolutionary line on
one planet only.

As an example of a universal solution, consider the immensely thick
legs of an elephant. That is surely a universal solution to the problem of supporting great bulk on a planet with a relatively high gravity. On the other
hand, the elephant also has a trunk, which looks like a contingent characteristic that's likely to occur only on Earth. So, universal versus parochial,
which is which?

Joints are universal, but knees and elbows, and five fingers are parochial.
The development of vertebrate gonads and their ducts is totally mixed up
with the kidneys and urinary ducts. So our reproductive behavior has all
kinds of parasitological hangups. It might well have been anticipated, then,
that an intelligent vertebrate social species would find much of its social
excitement in sexual guilts. Aliens that don't share this set of parochials are
unlikely to have faces like ours—eyes, for sure, but not nose, external ears,
teeth—and will not share our prurient interest in certain experiences. So there
is probably no alien pornography anywhere.

Certainly one of the best places to look for interesting alien physiology is
in the science fiction literature. Here are a couple of notable examples:

• *A Mission of Gravity:* This famous story by Hal Clement, written in
the 1950s, is set on the giant, rapidly rotating planet Mesklin. It has a variable
gravity of 600 times Earth's at the flattened poles and only 4 times Earth's at
the equator. The intelligent Mesklinites are like enormous centipedes, having
evolved from aquatic jet-propelled ancestors. The story revolves about an
odyssey to rescue a United Planets probe that crashed at Mesklin's south
pole. Barlennan, the hero, is an odd Mesklinite who is prepared to endure
enormous psychological stresses, such as being six inches off the ground and
having a heavy object suspended above him briefly, in order to learn new
things. Figure 6.2 depicts a typical Mesklinite.

Figure 6.2. A typical Mesklinite from *A Mission of Gravity*

• *Memoirs of a Spacewoman:* This story by Naomi Mitchison (sister of the famed geneticist J.B.S. Haldane) is about the Radiates, who live on the planet Lambda 771. They have evolved from a radial form of life somewhat like a terrestrial starfish. The Radiates have five retractable arms that are studded with delicate suckers used for grasping tools and artifacts. On the top of their bodies, Radiates have a ring of bright blue eyes that encircle the brain case. Their bodies range in size from a few centimeters to about one meter in diameter. A typical Radiate is shown in Figure 6.3.

The Radiates live in villages composed of long, low buildings, roofed but with open sides, much like the longhouses found in places like Borneo. They decorate their ceilings with plants and fungi that grow in spiral patterns. The Radiates do not think in dualities like on/off or right/wrong, but use instead a five-valued system of logic. At intervals, all members of a community will join in an interlocking, wheeling dance, with those on the outer edge attempting to move closer to the center; this dance has strong emotional overtones of closeness and community cohesion.

The Mesklinites and the Radiates are but two of a vast array of alien life-forms that have been invented and explored by the science fiction community. But what are the constraints, if any, imposed on alien behavior by their morphology? In particular, since intelligence can be considered to be a behavioral characteristic, can we justify the argument that alien beings

Figure 6.3. A Radiate from *Memoirs of a Spacewoman*

with nonhumanoid forms might still evolutionarily converge to a humanlike intelligence?

Behavior and Morphology

What is revealed by examining a creature's bodily form (morphology) and its behavior? An excellent example is the now-extinct pterosaur Pteranodon. Its morphology can be summarized as a boxlike rigid body, large swept-forward wings, a crested beak, and hollow bones. Essentially, it was a glider, probably capable of powered flight for brief periods. This creature, shown in Figure 6.4, seems to have soared over warm Cretaceous seas, spending much of its life aloft feeding on fish, which it caught by trailing its long beak in the water. The crest may have functioned to keep the beak facing in the direction of the wind, much like a weathervane.

Clearly, interpretations of the behavior of Pteranodon are closely linked to interpretations of the biological and evolutionary functions served by the various parts of its bodily form. So understanding a creature's behavior cannot be divorced from the character of its bodily form. But what about social behaviors? Soaring over ancient seas and snatching up fish is certainly inter-

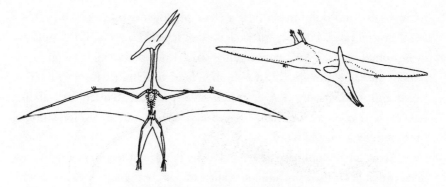

Figure 6.4. The pterosaur Pteranodon

esting. But this is not the kind of behavior that gives rise to a culture. Can cultural behaviors also be thought of as consequences of morphology?

It is obvious that a particular morphology restricts the actual bodily movements that a given organism can physically perform. But the exact sequence of movements performed in social behaviors is not immediately evident, though some might be implicated. The antlers of deer, the long horn of the narwhal, and the thickened skulls of rams, all used in mating contests between males, are a case in point. But for a given creature one cannot, in advance, generally predict the character its social behaviors would take. Given *only* the morphology of an organism, no reliable predictions can be made as to the actual manifestation of *any* of its social or communicative behaviors, the very behaviors we are most interested in vis-à-vis any putative alien life-form. So the one conclusion that can be drawn is that attributing *hypothetical* behaviors to *hypothetical* creatures with *hypothetical* bodily characteristics is fraught with deep theoretical difficulties.

A putative alien is a hypothetical creature. The physicist Philip Morrison has argued that the bodily convergence in tuna, ichthyosaur, and dolphin is a prime example of a convergent solution to a common problem among fish, reptile, and mammal. He goes on to claim that the evolutionary histories of these sea dwellers are essentially independent of one another, so why shouldn't we expect to see common problems constrain common solutions for aliens, as well?

But the anthropologist le gros Clark has pointed out that convergence in physical traits among creatures from separate phyletic groups could not be treated as signifying complete genetic independence. In fact, tuna, ichthyosaur, and dolphin do bear a close phylogenetic relationship: They are all vertebrates. Further, the skeletons of ichthyosaurs and dolphins are even homologous. Thus, convergence would not be unexpected in members of the same phylum. The forelimb of land vertebrates, from which dolphins and ichthyosaurs derive, initially five-toed and adapted for walking, has been readapted for many new functions: one-toed, tip-toe-running limb; flipper, arm, and hand.

From an evolutionary perspective, the important features of a putative alien is that (1) it shares no common ancestor, or common environment, with terrestrial creatures, and (2) it possesses a bodily form independently evolved, genetically unrelated, and morphologically distinct from that of humans—though it is posited to have a similar intelligence. So what we usually see put forth is a hypothetical creature with nonhumanoid morphology and humanoid intelligence. But since such an alien shares no common ancestor or common environment, how could it possibly respond to a common evolutionary opportunity? Due to the intimate relationship between bodily form and behavior, convergence to a similar intelligence in a totally different type of bodily form seems highly improbable. Let's spend a few pages examining this issue of alien intelligence in more detail.

The Mind of an Alien

A few years ago, I had a discussion with the eminent cosmologist Jim Hartle of the University of California at Santa Barbara, who told me that he and Murray Gell-Mann, my colleague at the Santa Fe Institute, were working on a theory that would show that any intelligent being anywhere in the galaxy would have to conceptualize the physical universe in exactly the same way. In other words, intelligent beings on the other side of the Milky Way would still think about the interaction of objects in terms of positions, velocities, forces, and all the other concepts we use here on Earth to describe such processes. Faced with such an amazing claim, I pressed Hartle for a prepublication copy of this work, thinking that if it were really true it would constitute one of the most significant philosophical, as well as psychological, discoveries of all time. But, alas, at last hearing Jim and Murray's work has still not reached the publication stage.

My conversation with Hartle raises the fascinating question of what psychologists call "cognitive universals." These are concepts that our civilization and any other technologically advanced extraterrestrial civilization can easily interpret. In short, it is just such universals that are necessary if we Earthlings and ETIs are to have a common basis for communication. Such concepts, if they indeed exist, are what would be needed for developing advanced technologies of the type that could send signals—or objects—across interstellar space. Since this notion of a cognitive universal is so important to the case for human communication with an ETI, let's be just a bit more specific about what it entails.

Cognitive universals are concepts that are easily interpretable by us humans and by ETI civilizations that (1) are at a technological level enabling them to have the capability of sending and receiving communications with other extraterrestrial civilizations, and (2) are able to place objects off the ETIs' home planet. The first requirement is a standard assumption of the SETI community, while the second is made to guarantee an operational utilization of the civilization's notion of geometry, at least in a local sense in its immediate spatial neighborhood.

Examples of cognitive universals seen in the research literature have been constructed out of mathematical, physical, and chemical concepts. Most common is the notion that elementary arithmetic is necessary for technological development and thus that various cognitive universals can be constructed from arithmetic structures. Reasoned arguments for the necessity of arithmetic for technological development are very difficult to find, however, and most investigators ultimately fall back upon an "I can't imagine how it could be otherwise" type of example.

As an illustration, it's often put forth that ETIs will have to have knowledge of the entire set of natural numbers (1, 2, 3 . . .), leading the investigator to claim that once the natural numbers are in hand, prime numbers won't be far behind. Thus follows a proposal for broadcasting the sequence of prime numbers as a kind of beacon to attract the attention of ETIs (as was the premise of Carl Sagan's book *Contact* and its subsequent hit film). But it's perfectly possible to imagine simple arithmetical calculations involving natural numbers that can be performed without having the general concept of "natural number." That is, without having the notion of the *entire set* of natural numbers.

These considerations show that more than calculative utility is needed to establish the universality of inductively generated numerical concepts like the natural numbers. For SETI, this means that we should be very wary of the use of concepts such as "prime number" or "the binary expansion of π" as a common basis for communication with ETIs. These inductively based numerical concepts ultimately derive from metalinguistic abilities, abilities whose universality is questionable in the extreme. It is simply not enough to say that integers are useful in various kinds of estimations and calculations, for these can take place without a concept corresponding to the totality of numbers.

Perhaps perceptual constancies, like the cognitive representation of physical space and the representation of physical intensity of objects, might serve the purpose of communication better. In these cases, there are only a few mathematical groups under which such representations remain unchanged—rotations, translations, reflections. On evolutionary grounds, for each of these types of groups there is much in common about the ETIs' psychological processing of the associated constancies. So these groups may be much more fertile places to look for cognitive universals than in the realm of arithmetic.

But what do we mean by an alien intelligence, anyway? Some have argued that such intelligences already exist here on our own planet, residing in the brains of deep-sea creatures like the squid, the octopus, the dolphin, and the whale. And if we cannot effectively communicate with these creatures with whom we share a long evolutionary history, how could we ever imagine communicating with an extraterrestrial that has evolved in a vastly different environment under very different circumstances? So let's have a look at the cephalopods like the octopus to see what can be learned about intelligence from these "alien" creatures.

What Is This Octopus Thinking?

In 1992, neuroscientists Graziano Fiorito and Pietro Scotto in Naples trained one group of octopuses to choose a red ball or a white one. Their goal was to study learning in the common octopus species *O. vulgaris*. The idea was to see if an octopus could learn a skill by observing the activities of other octopuses.

When the trained animals reliably approached one or the other ball, untrained octopuses were allowed to watch. When later presented with a choice of their own, the untrained animals not only selected the same ball more often throughout the five days of the trial, but also learned more quickly through observing than the original subjects had under classical Pavlovian conditioning.

The researchers concluded that "the rapid acquisition and the stability indicate that observational learning in *O. vulgaris* is a powerful mechanism of learning." The finding was astounding not least because observational learning is considered by some to be a preliminary step to conceptual thought.

But there are critics like Jean Boal of the University of Texas at Galveston who say that the question isn't what kind of complex learning cephalopods are capable of, but whether they're capable of it at all. First, she says, there's the problem of why the solitary octopus has evolved the capacity to learn by observing other octopuses. After all, octopuses are short-lived, solitary creatures that usually only meet once to copulate. So why do they need to learn by example? Supporters of the smart octopuses counter that other cephalopods form loose social groups, so the octopus may have inherited its penchant for observational learning from a common evolutionary ancestor.

And so the experiments continue. But until fresh results are in, the debate about the intelligence of the cephalopod will remain deadlocked. On the one side, skeptics like Boal predict that the cephalopod will prove nowhere near as intelligent as the earlier studies claim. On the other side, there are those who firmly believe it's only a matter of time before the depth and breadth of cephalopod psychology is finally revealed. Says Nathan Tubitz of the University of Oregon, "We have not yet come up with the right set of experiments to illuminate the intelligence these animals possess. The problem is the limitation of humans, rather than the limitation of the cephalopods." Perhaps we'll discover the same limitations when it comes to communication with ETIs.

Communication of any kind is going to involve some sort of language. So for the sake of argument, let's assume the Earthlings and the ETIs do have a sufficient set of cognitive universals to make communication at least feasible. What linguistic structure might be used as a bridge across this communication gap?

Talking the Talk

When American astronauts explored the Moon in the late 1960s and 1970s, it took one and a half seconds for the radio broadcasts from Earth to reach them and for their replies to be received back at NASA Mission Control. We could all hear the awkward delay in the conversations between the mission controllers and Neil Armstrong. When it comes to communication with ETIs, if that communication is by radio, which seems to be most likely, the round-trip conversation times are going to be measured in years—or even centuries—not seconds. The pauses between replies are going to be on the order of human lifetimes.

Of course, speaking with ETIs face-to-face would not pose such time-lag difficulties. But *any* attempt to communicate with aliens from other worlds will run up against three fundamental problems: (1) *Can* we talk? (2) How *will* we talk? (3) What will we talk *about*? Here let's focus just on the first two of these problems.

For the sake of argument—and because it's not an unreasonable assumption—let's take it as axiomatic that alien communication systems will be true languages. Thus, they will use small collections of elements like sounds or movements that are infused with an arbitrary set of meanings. Their language signals will have stable associations with their real-world experiences. They will pass on their language systems to succeeding generations by teaching and learning, but their infants will also be born with some innate ability to learn their alien mother tongues. Their language will bind both space and time, conveying information about their pasts and possible futures. Clearly, these requirements are highly anthropomorphic in character, but necessary assumptions if we humans are to make meaningful contact with the alien language systems.

But what points of contact will such an alien language have with a human language? Probably, not much, if the alien language uses some modality or delivery system unknown to us. For example, an alien language might be based on smells. Such a language most likely could not exist on Earth, since pheromones, the chemical compounds our noses detect and our brains interpret as odors, do not travel fast enough or far enough to be a good medium for language. On an alien planet, though, pheromones might be the perfect vehicle for linguistic information. But an alien who "speaks" using smells will not communicate much of anything on Earth, just as humans making auditory noises might not get very far on Planet X.

However, if we meet aliens who communicate using sounds or movements, then we may stand a chance of finding a way to translate their language into ours. At least we will share the same delivery system. This makes solving the second problem of how we will communicate a lot easier than it might be otherwise. We will first try to figure out the respective language symbols for the most basic things we have in common. Some claim that this requires us to begin with mathematics and numbers. But as we have seen in the preceding section, there are many potential difficulties with this approach, not the least of which is that it might be very difficult for me to explain to you my hike yesterday in the Sangre de Cristo Mountains using the Pythagorean Theorem! Let's consider just how different languages can be by looking at a couple of linguistic systems here on Earth.

Really Foreign Languages

Some years back, the popular comedy film *The Gods Must Be Crazy* had characters in it speaking a language that had numerous clicking sounds as part of its sound system. These clicks are so utterly different from any of the sounds in English that native English speakers have trouble even hearing them as speech. Chances are that when hearing a Bushman speak, your ear will interpret it as someone speaking a "normal" clickless language, with an unidentified separate source making clicking noises in the background.

In another direction, in many African and Asian languages the tone in which a word is pronounced is integral to its identity and drastically affects its meaning. You might object, saying, but we do similar things in English. For instance, I might say, "Come here!" in tones expressing urgency. But that's not the way a tone is used in, say, Chinese. There, a single word is pronounced in four tones—high level, low level, rising, and falling—that give it four completely unrelated meanings, as different as "spinach," "car," "house," and "friction."

All these variations, of course, are still custom-fitted to the specific type of sound-making machinery we have been given by nature, and which is peculiar to all human beings. What are the chances that a being evolving on a faraway planet in a vastly different environment will have a vocal system so close to ours that it can speak with sounds even as close to English as Arabic or Chinese?

Think about birds and whales. They simply don't have the same kind of vocal equipment humans do. Therefore, the sounds they make are too different to be mutually pronounceable, or for theirs to be represented with any accuracy by our spelling system. For example, bird guides often contain something like "Whee-wheeoo-titi-whee" to describe the call of a particular bird. This kind of transcription does such a poor job of conveying what the real thing sounds like that you're unlikely to recognize it when you hear it. The problem is that while the rhythm and intonation are approximately correct, the vowels and consonants aren't even vaguely there.

And what about aliens who don't even use sound as their medium of communication? We've already spoken about smells. But even here on Earth we have the dance of the honeybees or the high-pitched sonar of the bats (sound beyond the human range of perception).

The situation gets even more complicated when we consider that even in Earth-based languages the basic unit is not necessarily the word. So it's not out of the question to imagine an alien language with grammatical structure wildly different from anything we have seen here on Earth. Just as a small example, consider the Swahili language. It turns out that Swahili pronunciation and grammar are both quite easy for an English speaker to learn, because they are very logical and regular. But the language arranges things in a way completely new to an Indo-European speaker.

In English or Spanish or German, the basic unit is the word, which may be a verb or a noun of any of several types. Some of these words undergo changes to express things like verb tense (present or past) or noun case (subject or object). Usually these changes occur at the end of a word. In Swahili, the word is not quite so basic. Much of the grammar that must be learned is rules for *building* words by assembling smaller elements. Most of the elements that are added on to "foundation" blocks such as verb stems are added at the front, not at the back as in English or German. This makes using a dictionary without a good basic knowledge of grammar a bit harder, since the root you look up may be close to the end. The root of *kilichotutosha,* for example, is *tosha,* and you have to know what the prefixes *ki, li, cho,* and *tu* mean when attached to it in that order.

This is just one example of how differently languages can be constructed. Since a language like English is designed for a nervous system peculiar to humans, we can easily expect that beings from other places and other environments might use structures so profoundly alien that we will never fathom them. Here's another example of how this might happen, again even here on Earth.

The WIPP Marker

Just outside Carlsbad, New Mexico, the U.S. Department of Energy (DOE) has constructed a repository 2,100 feet below the desert floor, where over 6 million cubic feet of nuclear waste will be entombed. Since the half-life of plutonium is twenty-four thousand years, the agency must consider the prospect of people from the far-distant future poking around the site long after fences and guards have turned to dust and no one remembers what's there. Thus the need for a modern Stonehenge, durable yet also decipherable: a warning to the next few hundred generations to stay away. So here we have the problem of trying to communicate with "aliens," who are our own descendants—but in deep time, more than ten thousand years from now.

In 1992, the DOE asked two independent teams of experts—anthropologists, linguists, artists, science fiction writers, and materials scientists—to come up with ideas for a monument that would survive for ten millennia and still deliver a clear message. The two panels agreed on many things. The first was that to build a structure that could stand for ten thousand years was no problem. The great pyramid of Cheops is more than 90 percent intact after forty-six hundred years and probably has several millennia left. Modern materials scientists should be able to do at least as well.

On the other hand, building a marking system that will survive is a far trickier business. The danger comes not only from vandals but also from recyclers. Stonehenge is a good example, as in the Middle Ages many of its stones were carted away for use on other construction projects. The way to prevent recycling, according to the investigators, would be to build the structures of worthless material—local soil and rock, cut irregularly, so as to be hard to reuse.

Both panels also agreed that the structure should be "colossal." One reason is simply that a large structure survives longer. A more subtle reason, though, is that to cut through the vast amount of "memorabilia" that our civilization will leave to our descendants of ten thousand years from now, the waste-site markers will have to be conceived on a scale equivalent to that of the pyramids in order to grab the attention of overstimulated scholars of the future.

In the end, both panels came up with pretty much the same design for a marking system: massive earthen berms enclosing a field of monoliths inscribed with warnings in many languages. Figure 6.5 shows two examples offered by the panels, the Spike Field and the Landscape of Thorns. In both cases the idea is to produce a sense of dread and unease in people of the future, so that they will stay away from the waste site.

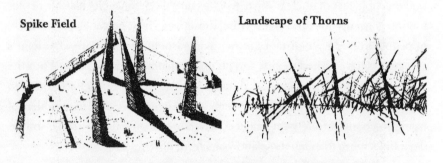

Spike Field   **Landscape of Thorns**

Figure 6.5. The Spike Field and the Landscape of Thorns

As the Waste Isolation Pilot Plant study indicates, pictures are one way of cutting through the communication barrier presented by communication systems based on sounds. Several SETI investigators have suggested pictorial messages as the best way to communicate with ETIs, and a couple of early efforts in this direction are discussed in *Paradigms Lost.* All of these attempts are based on coding a two-dimensional picture using black-and-white squares in a rectangular grid. The number of squares on the sides of the grid are taken to be prime numbers, on the assumption that ETIs will know and understand arithmetic and, hence, will know and understand the importance of the prime numbers (a dubious assumption discussed above).

Just to throw a bit of cold water on this pictogram idea of communication with ETIs, Figure 6.6 shows a page of the mysterious Voynich manuscript held at the Yale University Library. This strange document is a 232-page illuminated book written entirely in a cipher that has never been

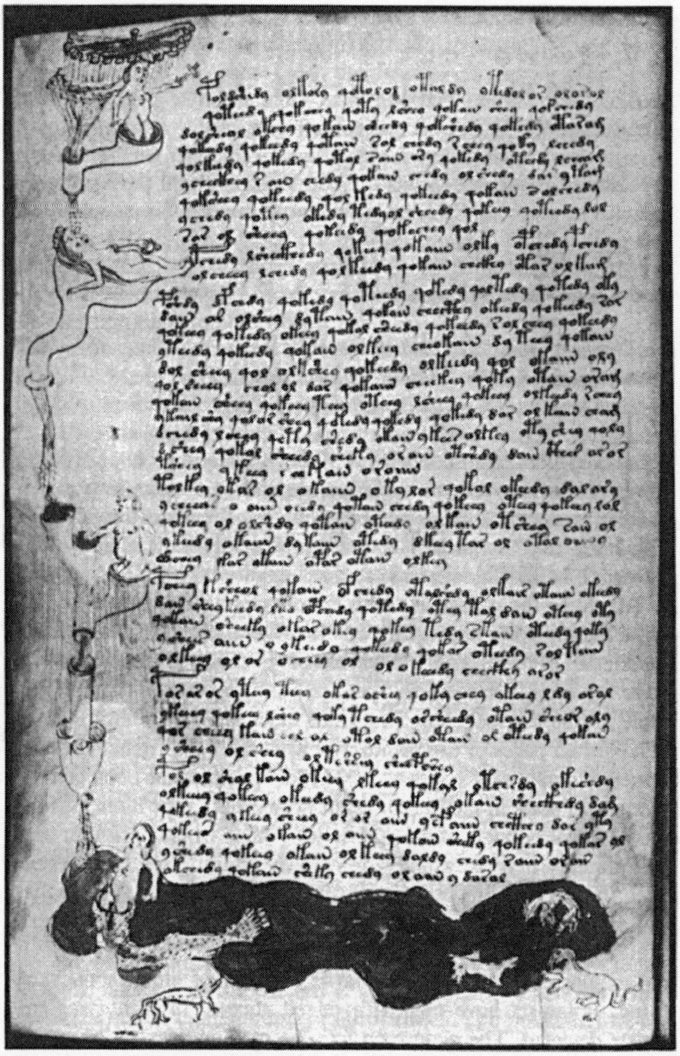

Figure 6.6. A sample page and symbols from the Voynich manuscript

decoded. Its author, subject matter, and meaning are complete mysteries. No one even knows what language the text would be in if it were deciphered. The pages show fanciful pictures of nude women, peculiar inventions, and nonexistent flora and fauna. Plans for weird plumbing show nymphets frolicking in sitz baths connected with branching elbow-macaroni pipes. In fact, the whole manuscript has the eerie feeling of a perfectly sensible book from some "second Earth" in another galaxy. So here we have an example of something created on Earth—but in a symbol system that no one can make sense of. Why do we think we could do better with an analogous manuscript from ETIs?

Much of what we've discussed about intelligence, communication, and language rests on the explicit—or implicit—anthropomorphic assumption that ETIs will be not too different from us. This is just one form of the Copernican Principle of Mediocrity, asserting that everything about us humans is pretty much typical of the way things are in the galaxy at large. Recently, Princeton physicist J. Richard Gott III proposed some new twists on this idea. Let's have a look at how his view of mediocrity applies to SETI.

According to Gott

In 1969, Richard Gott III took a summer holiday in Europe before continuing his studies in astrophysics at Princeton. During this trip he visited the Berlin Wall. As he stood looking at the wall and pondering the cold war, Gott wondered how long it would last. Having no special knowledge of East-West relations, he reasoned, first of all, that there was nothing special about the timing of his visit. In other words, he didn't come to see the wall being erected or demolished—he just happened to have a holiday and be standing there at a random moment during the wall's existence.

So, he argued to himself, there was a fifty-fifty chance that he was seeing the wall during the middle two quarters of its lifetime. If he was at the beginning of this interval, then one quarter of the wall's life had passed and three quarters remained. On the other hand, if he was there at the end of this interval, then three quarters had passed and only one quarter lay in the future. In this way, he calculated that there was a 50 percent chance the wall would last from ⅓ to 3 times as long as it already had. Since the wall was then 8 years old, Gott predicted that it would last anywhere from another 2⅔ years (8 times ⅓) to 24 years. As it happened, the Berlin Wall unexpectedly came crashing down in November 1989, just 20 years after Gott's visit—right in line with his prediction.

More than a decade later, Gott refined this Principle of Indifference, adopting the more standard scientific criterion that predictions should have at least a 95 percent chance of being correct instead of the 50 percent he used in the case of the Berlin Wall. This makes the numbers come out slightly different, but the underlying argument remains the same. If there is nothing special about your observation of something, then there is a 95 percent chance you are seeing it during the middle 95 percent of its observable lifetime, rather than during the first or last 2½ percent. At one extreme, then, the future is only ⅟39 as long; at the other end, it is 39 times as long. So with 95 percent certainty, the future longevity of whatever you observe is between ⅟39 and 39 times as long as its past.

As a test, Gott used this principle to predict the future longevities of the forty-four Broadway and off-Broadway plays and musicals running in New York in 1993, the time of publication of his article on the method in the British science weekly *Nature.* In each case, the future longevity was within a factor of 39 of the past longevity, just as predicted. This whole line of argument is strongly reminiscent of the Copernican Principle, which we discussed earlier. So let's see how the two connect, and the implication of Gott's method for SETI.

Copernicus, Gott, and SETI

Four centuries ago when Nicolaus Copernicus pointed out that the Earth revolved about the Sun rather than the other way around, he immediately displaced humanity from its privileged position at the very center of the universe. We now see the Earth as circling a very ordinary star in the suburbs of the Milky Way galaxy, a position that's completely undistinguished in every possible way. This leads to the Copernican Principle, which can be regarded as the supposition that one's location is unlikely to be special.

A good test of this principle came earlier this century, when astronomer Edwin Hubble observed approximately the same number of galaxies receding from Earth in all directions. So, at first glance, it seemed as if our galaxy was at the very center of the universe. But reasoning with the Copernican Principle, scientists concluded instead that the universe must look that same way to observers in every galaxy, as it would be presumptuous in the extreme to suppose the Milky Way galaxy is in any way special. But the Copernican Principle doesn't apply only to placement of galaxies in space—it works inside of time as well. Gott's Principle of Indifference is simply a way of attaching some numbers to the Copernican Principle and using it in time as well as in space.

What does this principle tell us about our own lifetime as a species? We have now been around for about 200,000 years. If there is nothing special about the present moment in time, then it is 95 percent certain that the future duration of *Homo sapiens* is between $\frac{1}{39}$ and 39 times this number. So we should last for at least another 5,100 years but less than 7.8 million years.

This estimate gives our species a likely total longevity of between 205,000 and 8 million years, numbers that are completely in line with those for other hominids and mammals. For instance, our ancestor *Homo erectus* lasted 1.6 million years, while Neanderthal Man survived 300,000 years. The average duration of mammalian species is 2 million years, while the great dinosaur *Tyrannosaurus rex* hung on for 2.5 million years.

Before continuing this discussion of Gott's Principle of Indifference, it should be noted that there are a number of contrarians who argue that the whole idea is nonsense. Steven Goodman of Johns Hopkins University notes that this principle has its origins at least as far back as Leibniz in the seventeenth century. In the 1920s, John Maynard Keynes stated, "No other formula in the alchemy of logic has exerted more astonishing powers. For it has established the existence of God from total ignorance, and it has measured with numerical precision the probability that the sun will rise tomorrow." In short, the principle says that if you know nothing about a specified number of possible outcomes, you can assign equal probability to each of them. This is exactly what Gott does when he assigns a probability of 2½ percent to each of the 40 segments of a hypothetical lifetime. I'll not go into the arguments put forth by Goodman and others against Gott's principle, but only refer the interested reader to source material for them listed in the "To Dig Deeper" section for this chapter.

Now what about the implications of this principle for SETI?

One of the standard Defense arguments against the existence of ETIs is the Fermi question: Where are they? This argument is based on the Copernican Principle that we are nothing special, so that there must exist ETIs far more technologically advanced than us. And these ETIs would certainly have begun colonization of the galaxy, which implies that we should certainly have seen some signs of them by now. But even though the Copernican Principle and the Gott Principle of Indifference are based on the same line of reasoning, their conclusions are quite different in this case.

Gott says that colonization is not important in the sense that galactic colonists and their descendants must not dominate the numbers of intelligent observers in the universe, since otherwise we Earthlings would be likely to be one. So this explains why we should not be surprised that we have not been colonized by ETIs. Assume that there are a billion habitable planets in the galaxy. A lengthy calculation leads to the conclusion that with 95 percent likelihood, the average longevity of a technological civilization is less than 12,100 years, and not more than 121 civilizations are now transmitting signals into outer space. These numbers suggest that there is *some* possibility that a radio search for other civilizations in our galaxy could be successful. But a targeted radio search of 1,000 nearby stars is not likely to succeed.

Gott concludes his article on the Principle of Indifference with a sobering conclusion:

> The odds are against our colonizing the Galaxy and surviving to the far future, not because these things are intrinsically beyond our capabilities, but because living things usually do not live up to their maximum potential. Intelligence is a capability which gives us in principle a vast potential if we could only use it to its maximum capacity, but so does the ability to lay 30 million eggs as the ocean sunfish does. We should know that to succeed the way we would like, we will have to do something truly remarkable (such as colonizing space), something which most intelligent species do not do.

Let's now leave Gott and his Principle of Indifference and have a long look at exactly how the search for ETI is being carried on today in various centers around the world.

The Search Goes On

On August 15, 1977, the "Big Ear" radio telescope at Ohio State University was tuned to the 1,420-megahertz frequency at which hydrogen atoms of interstellar space broadcast radio waves like miniature radio transmitters. In those long-ago days, computer technology was in its infancy and so the fluctuating voltages picked up by the telescope could not be stored electronically but had to be printed out and later scanned by SETI volunteers. One such volunteer was Jerry Ehman, now a professor at Ohio State.

As Ehman worked his way through the printouts spewed forth by the lab's printer that day, he got the shock of his life. At one point, the voltages jumped clear through the roof. A powerful radio signal had been recorded by the telescope, coming from the direction of Sagittarius and had lasted thirty-seven seconds. Ehman scrawled "Wow!" in the margin of the printout. To this day, this unexplained burst of radio noise is universally remembered in the SETI community as the Wow signal.

All that Ehman and John Krauss, director of the Ohio State observatory, were able to determine is that the WOW signal did not come from Earth or from a satellite. "The source of the emission must remain open," says Ehman. Arthur C. Clarke, the noted science fiction writer, agrees. "God knows what it was," he says.

By almost common consensus, the Ohio State WOW signal is the way that we will first make contact with ETI. Romantic visions of alien abductions, landings of strange spacecraft on the White House lawn, or *Star Trek*-like explorations where no man has gone before aside, the overwhelmingly likely way to cover the vast distances of interstellar space is via signals in the electromagnetic spectrum, most likely radio transmissions. So the vast majority of the meager resources available today for SETI are devoted to listening for signals from alien civilizations presumed to be far more advanced than our own.

One good way to get a handle on how to listen and what we might expect to hear is to see what kind of emissions a satellite flying by the Earth would receive. Then we would look to see if these emissions would be sufficient for an intelligent observer to conclude the existence of sentient life-forms on our own planet. Just such an experiment was conducted in 1990.

The *Galileo* Flyby

Galileo is a spacecraft that was designed to pass by Jupiter in December 1995. Through a fortuitous combination of circumstances, the spacecraft's trajectory also enabled it to pass close by the Earth in December 1990. As *Galileo*'s instruments were not designed for an Earth encounter, this led to the arrangement of a fascinating control experiment: a search for life on Earth with a typical modern planetary space probe. Using such instruments, could one conclude the existence of life on Earth?

Galileo's closest approach to Earth was on December 8, 1990, about 960 kilometers over the Caribbean Sea. At that time the Earth was approached from the night side. In the search for life on Earth, data was collected in the near-infrared and ultraviolet parts of the spectrum, as well as from the solid-state imaging system and the plasma wave spectrometer. Investigators did not assume known properties of life on Earth, but instead attempted to deduce their conclusions from the *Galileo* data and first principles alone. For example, a necessary condition for the presence of life is a marked departure from thermodynamic equilibrium. Once candidate disequilibria are identified, alternative explanations must be eliminated. Life is the explanation of last resort. *Galileo* found such profound departures from equilibrium that the presence of life seems the most probable cause. So what did the investigators find in this benchmark probe for the existence of life on Earth?

From the flyby, an observer otherwise unfamiliar with the Earth would be able to draw several conclusions. First, the planet is covered with large amounts of water present as vapor, as snow and ice, and as oceans. So if any living forms exist, they are plausibly water-based. There is so much molecular oxygen in the atmosphere as to cast doubt on the proposition that ultraviolet photodissociation of water vapor and the escape of hydrogen is its source. An alternative explanation is the biologically mediated photodissociation of water by visible light as the first step in photosynthesis by living organisms, namely, plants.

Methane was detected by *Galileo* some 140 orders of magnitude higher than its thermodynamic equilibrium value in such an oxygen-rich atmosphere. Only biological processes are likely to generate so large a disparity.

Several percent of the land surface area—entirely in Australia and Antarctica—was imaged at a resolution of a few kilometers. No unambiguous signs of a technologically based civilization were found. But narrow-band pulsed amplitude-modulated radio transmissions above the plasma frequency strongly suggests the presence of such a civilization. It's interesting to note that most of the evidence uncovered by *Galileo* would have been discovered by a similar spacecraft as long ago as 2 billion years—with the exception of the radio transmissions that could not have been detected before this century.

Unfortunately, while the *Galileo* mission established the ability of flyby spacecraft to detect life at various stages of evolutionary development on other worlds, at least in our own solar system, we are not yet sending space probes to the other side of the galaxy. So let's have a quick look at some of the sorts of listening experiments going on to find the next WOW signal—the type of space "probes" being sent to *our* side of the galaxy.

Who's Listening?

In 1993, the U.S. Congress canceled funding for both the Superconducting Supercollider and for SETI. At that point, an international group of radio and radio astronomy enthusiasts formed the SETI League, an organization dedicated to carrying on where the professionals left off. From their backyards around the world, they point satellite dishes at the sky and dream of the day when one of them will catch an ETI phoning Earth.

The SETI League's ambitious goal is to have five thousand amateur stations up and running by 2001. This is part of Project Argus, named after the giant of Greek mythology who had a hundred eyes looking in different directions. Joining Project Argus involves buying a few pieces of relatively cheap equipment: a satellite dish, a microwave receiver, and a computer with digital signal-processing software, at a total cost of less than $7,000. A typical Argus-type setup is shown in Figure 6.7. For readers interested in finding out more about this effort, the URL for contacting the SETI League is listed in the "To Dig Deeper" notes and references for this section.

A complementary listening effort aimed at involving the general public with SETI is being put together by University of Washington astronomer Woody Sullivan. This project, termed SETI@home by Sullivan, involves harnessing unused computer capability in people's home computers to search the data coming from SETI radio telescopes for an ETI signal. The SETI@home project distributes data from the telescopes, as well as software to parse patterns within that data, to anyone who asks for it via the Internet. The software runs automatically when the participating machine is in screen-saver mode. After a chunk of data has been analyzed for patterns of repetitive signals—the supposed hallmark of intelligent life—a report is automatically sent to the SETI@home server located at the University of California, Berkeley.

The source data sent to participants for analysis is collected by Serendip IV,

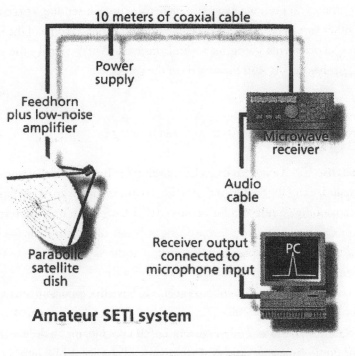

Figure 6.7. A typical amateur SETI system

a receiver that piggybacks on the 300-meter radio telescope at Arecibo, Puerto Rico. This receiver started scanning nearly 168 million radio channels in 1997. Sullivan estimates that if fifty thousand participants sign on to SETI@home, the computing power will be equivalent to a substantial fraction of a typical supercomputer but will cost only about half a million dollars. What a bargain! Both Argus and SETI@home are amateur setups. What have the professionals been doing?

SETI in the 1990s

During the past decade, there have been five major SETI programs carried out at observatories in the United States, as well as several smaller projects at other places around the world. For the sake of comparison, Table 6.4 gives a short summary of the American-based searches in the first half of the 1990s.

PROJECT	CHANNELS SEARCHED (MILLIONS)	ANTENNA DIAMETER (METERS)	SEARCH FREQUENCY (GIGAHERTZ)	FREQUENCY RESOLUTION (HERTZ)
Microwave Observing Project (NASA Ames, 1993)	29	70, 305	1 to 3	14, 28
Microwave Observing Project (NASA JPL, 1996)	16	34	1 to 10	30
Serendip III	106	305	1 to 10	0.6, 600
Project META (Planetary Society, 1994)	160	26, 30	1.4 to 1.7	0.5
Ohio State University (1993)	10	53	1.4 to 1.7	1

Table 6.4. American SETI searches in the early to mid-1990s

Naturally, these and other searches of the skies go on—even without the help of the federal government. Proponents of the existence of ETI firmly believe that one of these days another WOW signal is going to turn up in one of these telescopes, ushering in a new age of our awareness of our position in the universe. For the sake of discussion, suppose that contact comes tomorrow. Suppose you are one of Project Argus's listeners and you receive a signal from the stars. What happens next?

Contact!

The pulsar CT-102 was observed by Soviet radio astronomers in 1965. At that time, the regular radio emissions from the rapidly rotating neutron star were interpreted as being signals from an intelligent civilization, and were so reported by the Soviet news agency Tass. This announcement was front-page news around the world. Later, to its credit, the Soviet astronomical community rebuked Tass for this interpretation when it was discovered that these "signals" were the result of natural processes. But the episode illustrates how ill prepared the world's media is for dealing with first contact with ETIs.

As the foregoing discussion amply shows, it is really impossible to determine how likely we are to discover electromagnetic evidence of ETI. We have only theory and speculation to guide us. Several SETI scientists have stated that if we are so fortunate to be successful in this search, we can cross the bridge of how to announce and deal with the discovery of ETIs when we come to it. But many feel that advance preparation to cross that bridge of success is essential even though the likelihood of that crossing may be small.

In 1989, noted SETI scientist Donald Tarter conducted an international survey of individuals from the radio astronomy community and the science media, asking their views on the relative importance of SETI. On a scale of 10, SETI ranked 8.2 in terms of importance among the science media and 9.0 among SETI researchers. Over 90 percent of the science media endorsed the idea that there should be some designated body of scientists available to help interpret and analyze any announcement of the discovery of ETIs.

The problem of dealing with a signal from the deep are twofold: verification and interpretation. The problems of signal verification are mostly technical, involving primarily ruling out all possibilities for the source of the signal other than ETIs. The really hard questions arise when it comes to interpreting the signal. These problems are mostly psychological and sociological. Given the enormous public impact of a verified ETI signal, it seems only prudent to start now to develop procedures to analyze the possible content of such a signal, as well as developing some understanding of possible public reaction based on various possible signal contents. Let's look a little deeper at the characteristics of such an ETI signal, in order to understand better the complexities of verification and interpretation.

Characteristics of an ETI Signal

In his analysis of the problem of preparing the public for an ETI signal, Donald Tarter has identified five characteristics such a signal would possess. They are (1) certainty, (2) clarity, (3) sensory accessibility, (4) activity level, and (5) intentionality. Let's take a harder look at each.

• *Certainty:* How certain are we that the signal is really from an ETI? The Ohio State WOW signal is an example of the type of ambiguity that surrounds the question of certainty of the signal. False alarms like the Russian announcement concerning CT-102 are another. In envisaging an ETI discovery, we often imagine that it will be a discrete and certain announcement. But it's likely that this will not be the case. At whatever time an individual or agency thinks it has sufficiently strong evidence to go public with the story, there will probably be some level of scientific disagreement or skepticism about the announcement. So the fate of the competing interpretations that ensue will determine how the story plays out in terms of public impact.

If a scenario of gradually increasing certainty is what actually happens, then we might expect the exploitation potential of the event to be reduced. Potential exploiters would be more hesitant to make substantial investments in "ET hype" if there was initial uncertainty as to its reality. In fact, it might be an effective policy to keep the public regularly informed about promising—but unconfirmed—signals. Presently, such signals are kept within the astronomical community. But the gradual release of these interesting, but unconfirmed, signals might allow the public to become accustomed to the uncertainties surrounding ETI detection.

• *Clarity:* The search for meaning in a verified ETI signal will be of primary importance for both amateurs and professionals. Everybody will be offering their own interpretation, and the clarity with which the meaning can be deciphered will be a factor that greatly influences the ultimate social impact of an ETI discovery.

Even if we receive a complete ETI message and not just a fragment, most observers feel that it will be difficult if not impossible to fully understand the meaning of an ETI signal. Communication between entities with completely different origins and evolutionary experiences poses enormous barriers to a complete understanding of whatever signal we might receive.

A high-certainty, high-ambiguity event would greatly stimulate the "interpretation industry." Such a message would maximize the opportunity for malicious exploitation, as the public would be distracted from the scientific attempts to figure out what the message really means by the sideshow of events centering on those who claim to know the meaning or claim that the SETI community and/or the government knows the meaning and will not divulge it. In this scenario, there are enormous opportunities for rumor, exploitation, and fear.

In light of the possibilities for mis- and disinformation, a major order of business should be the development of policies that establish a relationship of trust, openness, and high credibility among the SETI community and the media and the public. A 1989 survey of the international science media indicated that observers from all the major cultural clusters (Western, Middle Eastern, sub-Saharan African, Far Eastern) thought it likely that there could be strong reactions from the more fundamentalist religious believers in their communities. Ways must be examined to communicate accurate, nonagitating information throughout the varying cultural regions of the world. So, there is the difficulty in communicating not only with ETIs but also with different cultures right here on Earth.

• *Sensory accessibility:* A message from ETIs might be in any form, from an audible code to visual signals. As the film *Contact* made clear, it's even possible that the message could contain instructions for how to make some type of receiver for an alien sensory encounter of taste, touch, or smell; or even how to construct an alien artifact.

It's likely that the social impact of the message would increase in direct proportion to the ways in which the alien message could be sensed. The intensity of interest would grow as sensory access increased.

• *Activity level:* An ETI message could vary from a simple passive greeting to a variety of requests or commands for us to take some dramatic form of action in response to the message. At the most passive level of communication, there might be an audio signal or an audiovisual signal much like the 1974 Arecibo message, humanity's first intentional message to space (see *Paradigms Lost,* Chapter 6, for a picture of this message). If the message were active, however, it could range from a request to "keep listening" to a command to make global adjustments in conformity with our new status as a cosmically communicating civilization.

Clearly, the social and psychological impact of the message will vary directly with its activity level. Even the simplest directives to keep listening might meet some resistance from the more xenophobic elements of Earth's societies. Requests for Earthlings to build something, do something to avoid danger, or change our ways entirely would, naturally, be the most socially controversial of all.

The wide range of possibilities for an active message demonstrates again the need for much forethought, planning, and effective public relations on the part of the SETI community.

• *Intentionality:* To what degree is the ETI message directed specifically to us, the inhabitants of Earth? How specifically are the aliens interested in us? This is the question of intentionality. Most likely, the intentionality of the message would be closely related to its meaning.

An unintentional message might contain the radio, radar, or television emissions from another civilization. Such discoveries would be totally unintentional on the part of the sending civilization. They are not designed for our eyes and ears.

On the other hand, the sending civilization might be scanning the galaxy for emerging civilizations. In that case the intentionality is not specific but general. They are looking for beings *like* us, but not us particularly.

Finally, it's not out of the question that we are the specific target of their transmission. Perhaps the sending civilization knows we are here; perhaps they have even visited us at some time in the remote past and plan to visit again; perhaps we have been discreetly monitored in our development. All these possibilities increase the likelihood that the social and psychological impact of the discovery would be very high. This could be unsettling, to say the least.

This typology of signal characteristics shows the possibilities of very complex social and psychological reactions to the receipt of an ETI message. It also illustrates the need for the international SETI community to begin to take steps to ensure the smoothest possible social assimilation of the fact that there are other intelligences "out there."

The typology can also serve as a tool for policy analysis. It provides us with a way to think through the problem of signal detection. We can construct complex scenarios based on variations of the characteristics of the ETI signal, and start to examine the social consequences of such a signal.

In closing this discussion of first contact, it's worth noting that a draft protocol for activities following the detection of a signal from ETIs has already been formulated by Allan Goodman of the School of Foreign Service at Georgetown University in Washington, D.C. Goodman notes that the principal operational problem is to find an appropriate and effective international consultative body to which the discovery of ETI signals and plans for analyzing and responding to them could be referred. He recommends establishment of a consultative organization that would be chartered under the auspices of the International Astronautical Union or another appropriate international scientific body.

Now let's summarize what's happened in the last decade insofar as SETI is concerned and try to come to some judgment on the cases for the Prosecution and the Defense.

THE APPEAL:
SUMMARY ARGUMENTS

By far, the most important development in the SETI game in the past decade has been the discovery of numerous planetary systems circling stars in other parts of the galaxy. The pace at which these discoveries are being made, together with the increased resolution of the findings, promises major advances in filling in the second term of the Drake equation, n_e, the fraction of stars having planetary systems with an Earthlike planet. Unfortunately, while a major advance for planetologists, this development still leaves us very far away from a good estimate for N, the number of communicating

EVIDENCE	INVESTIGATOR(S)	EVIDENCE FAVORS
Planetary systems	Mayor and Queloz, Wolszczan	Prosecution
Habitable zone	Levison	Prosecution
Form≠function	le gros Clark	Defense
Cognitive universals	Narens	Defense
Cephalopod intelligence	Boal	Defense
Cephalopod intelligence	Fiorito and Scotto/Tubitz	Prosecution
Language peculiarities	many	Defense
Principle of Indifference	Gott	Prosecution
Principle of Indifference	Goodman	Defense

Table 6.5. The evidence

ETI civilizations. And as we noted in *Paradigms Lost* in 1989, the difficulty in taking the Prosecution's case seriously does not lie in the astrophysical or geophysical terms in the Drake equation; the problems are with the remaining five terms, none of which has been improved upon one whit in the last decade. Table 6.5 summarizes the main new evidence, pro and con.

THE DECISION:
APPEAL DENIED

While the confirmed existence of planetary systems helps the Prosecution's case in some sense, it really doesn't help all that much in answering the question: Can we *communicate* with ETIs? This is where I drew the line in 1989 and it's where I draw the line today. Not a single shred of new evidence has been provided in the past decade to suggest that we will be able to meaningfully understand a signal from ETIs—even as being a signal from ETIs—let alone engage in any sort of meaningful dialogue with them. So until that evidence is forthcoming, the decision remains as it was a decade ago in favor of the case for the Defense.

SEVEN

The Way the

World Isn't

Claim: There Exists No Objective Reality

Independent of an Observer

BACKGROUND

A Polaroid Photo or a
Game of Twenty Questions?

It's always a bit spooky to watch a picture emerge from an exposed Polaroid snapshot, say of your son's or daughter's birthday party. Initially, there is nothing to be seen other than a uniformly gray piece of plastic. But as time unfolds, colors, shadings, and patterns slowly emerge, until after a minute or so a full-fledged picture appears of kids running around in funny hats with their faces smeared with chocolate ice cream. Suppose someone asked you if the picture was really there on the exposed film, and was only made manifest by the development process? Or was there no picture at all until the chemical processing took place? This really isn't much of a question for anyone who understands even the rudiments of photography, and we'd immediately answer that the picture of the birthday party really was encoded in the photosensitive chemicals of the film. We just couldn't see it until the development process acted on the chemicals to create something that our human eye could observe as a picture. Now let's change the situation.

Suppose you're at an adult birthday party, and the host asks everyone to participate in a strange variant of the familiar game of twenty questions. What's strange about this version of the game is that instead of having a target word fixed in advance, which the questions are trying to elucidate, there is no target word at all. Rather, as the contestant asks questions, the respondents must each give an answer, yes or no, that is completely consistent with all the answers given previously. In this fashion, at the end of twenty questions the contestant must produce some item—animal, vegetable, or mineral—that is consistent with all the answers given up to that point.

These two situations—the Polaroid picture and the target word in twenty questions—embody perfectly the two competing schools of thought in the quantum reality game. One school, call it the Realists, says that physical attributes of particles, things like position, velocity, and spin, are present *at all times*. But we only see the values of these attributes when we make a measurement on the particle, a measurement designed to bring out one attribute like velocity. So the Realists are like the Polaroid camera in the above analogy: The picture (read, attribute) is present on the film at all times. The film just needs to be processed (read, measured) in order to see it.

The loyal opposition, call it the Romantics, says no. Rather, its view is that attributes like velocity simply *do not exist* until we make a measurement. At that time, the attribute is brought into existence by the nature of the measurement apparatus, which is designed to literally *create* that attribute. Thus, the Romantics are analogous to people playing twenty questions when no target word has been selected in advance. The word (read, attribute) emerges as a result of the questioning (read, measurement/observation).

So these are the two versions of reality that we'll be exploring in this chapter. In short, we want to know, How real is the "real world"?

How Can It Be That Way?

In 1801, the English gentleman-scientist Thomas Young demonstrated the wave nature of light by an ingenious experiment. He made two thin, closely spaced vertical slits in a black curtain, and then allowed sunlight to pass through them and onto a white wall opposite the curtain. What he saw on the wall was a pattern of alternating black and white vertical stripes, which we now term an *interference* pattern. Young interpreted the dark stripes as the places where the crests of light waves passing through one of the slits met troughs of waves passing through the other slit, thus canceling each other out. The bright lines, on the other hand, he saw as places where a wave crest from one slit met a crest from the other slit and where trough met trough, thus mutually reinforcing each other's amplitudes. This famous experiment has served for over two centuries as proof that water, X rays, sound, and TV signals all display wavelike behavior.

Quite amazingly, the very same experiment can also be carried out with electrons—and with exactly the same results. The late quantum physicist Richard P. Feynman termed the interpretation of this *double-slit experiment* "the *only* mystery of quantum mechanics." The strangeness of the double-slit experiment with electrons is that when both slits are open, the pattern observed on, say, a phosphorescent plate like a television screen behind the slits is wholly like the one observed for light—and completely unlike what one would see if you threw balls at a picket fence with two slats missing. But if one of the slits is blocked, the pattern observed on the screen is identical to what one would see if the electrons were indeed just like baseballs. So that's the puzzle. How can an electron be both a particle and a wave? Put another way, how can the electron be localized in space and time like a baseball, while at the same time be smeared out over space and time like a water wave? This is the mystery of the quantum world. According to Feynman:

> I think it is safe to say that no one understands quantum mechanics. Do not keep saying to yourself, if you can possibly avoid it, "but how can it be like that?" because you will go down the drain into a blind alley from which nobody has yet escaped. Nobody knows how it can be like that.

The overall setup and results of the double-slit experiment are depicted in Figure 7.1a–c.

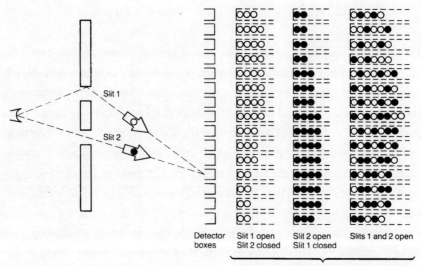

Figure 7.1a. The double-slit experiment with bullets

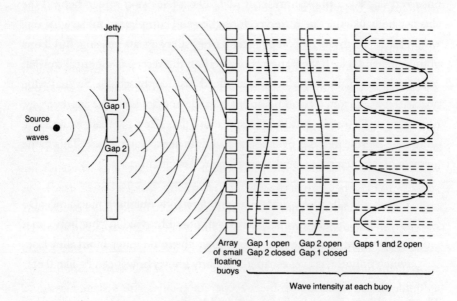

Figure 7.1b. The double-slit experiment with water waves

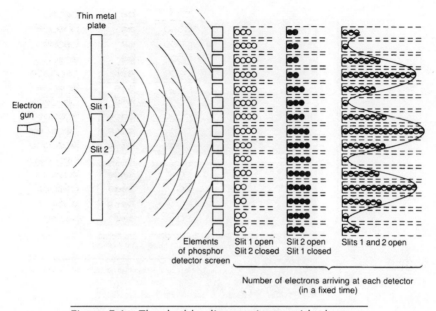

Figure 7.1c. The double-slit experiment with electrons

The conventional wisdom answer to Feynman's question is to associate a wave with the electron. But not just any old kind of wave; rather, it is a mathematical wave that is the solution of an equation called the *Schrödinger wave equation,* named after the great Austrian quantum physicist Erwin Schrödinger, who dreamt it up in 1929. This wave function describes the state of the electron at all times *when it is not being observed.* So when the slits are open, the wave function produces just the kind of interference pattern of troughs and crests that we see with water waves. One might interpret this pattern as the probability of finding the electron at a particular location when an observation of its position is actually made. In this view, the dark regions are where the electron is more likely to be found, the light regions where it's not so likely to be. Then, when the electron is actually observed (measured), this wave function magically collapses to the one particular point in space where the electron is actually found. To see that this is not just an artifact of an overly active mathematical imagination, Japanese physicists actually carried out an experiment in which single electrons were shot from an electron gun through a double slit and their appearance registered on a detector screen. Figure 7.2 shows what the screen looked like after the arrival of 10, 100, 3,000, and more electrons. What we see is that in this experiment, the wavelike interference

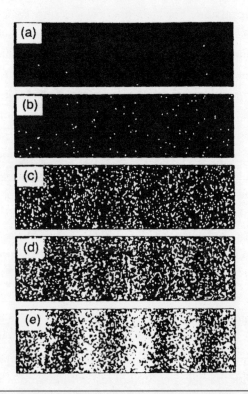

Figure 7.2. Buildup of an interference pattern for (a) 10, (b) 100, (c) 3,000, and (d)–(e) more electrons

pattern builds up, one electron at a time, as the number of electrons becomes large. So the phenomenon is not just mathematical; it's real.

This curious behavior in which the electron is a wave until it is detected, thereafter a particle, constitutes one of the major problems in quantum theory, the so-called Measurement Problem. We'll return to it shortly.

A lot of good physicists starting with Einstein have been both bewildered and unsatisfied by this kind of answer, which goes under the general rubric of the Complementarity Principle. Essentially, this principle states that quantum theory presents the observer with pairs of observables (or properties) of a quantum object that seem incompatible yet are both necessary for a complete view of the object. Knowledge of one observable hides knowledge of the other. In the case of the double-slit experiment, these complementary observables are the electron's position and velocity (technically, its momentum, which is the product of the electron's velocity and mass). Another pair of such dual observables are energy and time. The Heisenberg Uncertainty Principle tells us that if we want to measure one observable of such a pair, say, the position of the electron, very accurately, then we must accept a very inaccurate estimate of the value of the other observable, in this case the electron's velocity. The usual argument—which we'll see later is at best misleading, if not totally wrong—is that the more accurately you determine the electron's position, the more you disturb its speed and direction of motion. To be slightly more precise, two observables are "complementary" if precise knowledge of one of them implies that all possible outcomes of measurements of the other are equally likely.

The Uncertainty Principle is the refuge behind which physicists traditionally hide to explain how the interference pattern created by the unmeasured electron gets wiped out when an observation is actually made. But that is not the whole truth, and neither tradition, repetition, nor the authority of Feynman can make it so. What modern experiments show is that the interference pattern disappears even when the Uncertainty Principle does not come into play. We'll see why later. For now, let's return for a moment to the Measurement Problem.

We noted earlier that the complete state of the electron (all possible information needed to compute its position and velocity at any time) is contained in its wave function. The question that arises is how the wave function, a mathematical object described by the Schrödinger equation, makes contact with the purely physical properties of a particle's position and velocity. Two different worlds are being mixed here, one being the world of abstract mathematical objects and relationships to which the wave function belongs, the other being the physical world of material objects like electrons, billiard balls, and planets. Why should these two worlds have anything whatsoever to do with each other? And, in particular, how does the act of measurement, something carried out in the physical world, cause the collapse (or degeneration) of a purely mathematical construct, the wave function of the electron? This is the Quantum Measurement Problem. To state it more formally:

Quantum Measurement Problem

At exactly what point in the measurement of the electron's position does the wave function describing the probability of the electron's position "collapse"?

How does the act of observing the electron's position collapse the set of likelihoods into a single reality?

The Measurement Problem suggests that there is something "unreal" about properties like position, momentum, spin, and energy until we actually measure them with some type of observing apparatus. In other words, a property is not really a property until it is an *observed* property. Einstein hated this idea. He believed that observation plays no part in determining reality. Things are as they are, he might have said, whether you look at them or not. As he once remarked to his colleague Pascual Jordan, "Do you really think that the Moon exists only when you look at it?" In his view, it was the job of science to go beyond mere surface appearances and to describe and understand the nature of this objective, independent-of-human-affairs, rock-bottom kind of physical reality.

The nature of reality leads to the second Big Question of quantum theory, the so-called Quantum Interpretation Problem. In formal terms, it can be stated as follows:

The Quantum Interpretation Problem

What is the true "nature" of an unmeasured quantum object?

The best way to see what we mean by the term *nature* is to examine what might be called the orthodox and the reactionary schools of thought on the Interpretation Problem.

Orthodox View

1. The wave function gives a complete description of any *single* quantum object.

2. All quantum objects represented by the same wave function are physically identical.

3. The information an observer lacks about an unmeasured object is simply not there to be known.

4. The observed differences between identical unmeasured objects are due to inherent, that is, quantum, randomness in the objects.

Reactionary View

1. The wave function gives only a statistical description of an *ensemble* of quantum objects, hence a necessarily incomplete description of any single such object.

2. Quantum objects represented by the same wave function may not be physically identical.

3. The observer's ignorance about the attributes of an unmeasured object is due to the effect of certain "hidden" variables, which quantum theory conceals from view.

4. Objects with the same wave function may show differences upon observation because they were physically different before the measurement.

Those holding to the reactionary creed are often called *hidden variables* theorists for the obvious reason that they cling to a classical view of reality. Their credo is that once the properties and values of these hidden variables are known, then all the uncertainty about the values of attributes will fade away, and the quantum object will be seen as no different from a classical particle like a billiard ball or a baseball. The primary motivation for this view of reality is the desire to avoid somehow elevating the measurement process to a privileged position among the myriad physical actions that the universe might allow.

The key assumption separating the Reactionaries from the Conservatives is the second point on each list: the issue of whether or not there is a one-to-one correspondence between the grass-roots physical reality of dynamic attributes for objects, and the hard-to-get-your-hands-on mathematical reality of wave functions.

THE LOWER-COURT VERDICT

As with every Big Question in science and in life, there are myriad opinions and positions on these two central problems of quantum theory. A decade ago the competing positions looked like those in Tables 7.1 and 7.2.

THERE IS NO OBJECTIVE REALITY!

PROMOTER	ARGUMENT
Bohr (Copenhagen Interpretation)	overall measurement situation
von Neumann, Wigner (Consciousness Interpretation)	consciousness determines reality
Wheeler (Austin Interpretation)	measurement option
Heisenberg (Duplex Interpretation)	*potentia* and actuality
Everett, Deutsch (Many-Worlds Interpretation)	every world is a reality

Table 7.1. Summary arguments for the Prosecution

A SINGLE, OBSERVER-INDEPENDENT
REALITY MAY EXIST!

PROMOTER	ARGUMENT
Einstein (naive realist)	Newtonian reality is real
von Neumann, Finkelstein (quantum logic)	nondistributive logic
Bohm, Bell (quantum potential)	pilot-wave theory
Cramer (transactional events)	advanced and retarded waves

Table 7.2. Summary arguments for the Defense

The details of these various positions are well chronicled in Chapter 7 of *Paradigms Lost* so I'll waste no time reelaborating them here. However, for future reference it is worth briefly recalling the essential aspects of the most important ones.

The Copenhagen Interpretation, which is still the conventional wisdom among most physicists, asserts that it is the overall measurement situation that determines the nature and value of the properties that we can observe. For instance, if you set up a measurement situation designed to observe the position of the electron, then that situation brings into existence the electron's position while at the same time suppressing your ability to "see" the conjugate observable, the electron's momentum. The von Neumann view focuses on the interface between the measuring apparatus and the object being observed. To von Neumann's thinking, it is the observer's consciousness that creates the property that is measured. The other important argument is the rather radical many-worlds view of Hugh Everett III. Here the argument is that there is a separate reality for every possible value that a measurement can take. So, for instance, in one world the electron's position is 5 centimeters from the edge of a box containing it, in another world it is 5.1 centimeters from the edge, while in still a third universe the electron is on the opposite side of the box. For many-worlders, as soon as the measurement is taken, one universe is singled out, which in turn then branches itself into myriad possibilities for the next measurement.

On the opposite side of the fence are the realists, like Einstein and the late David Bohm, both of whom believed that quantum objects had definite properties at all times—whether or not they were being observed. Einstein's position is the one that most closely accords with our everyday sense of things, namely, that an electron is no different from the Moon: It has a definite position, velocity, spin, charge, energy . . . at each instance, and all we do when we measure it is see what the values of these quantities happen to be at the moment of measurement. David Bohm also held to this view, although without invoking any of the sorts of hidden variables advocated by Einstein.

The key theoretical idea that Bohm based his approach upon was the notion of a *pilot wave*. This idea had been introduced in the 1920s by de Broglie but was quickly laughed out of court by the Copenhagenists in view of what looked to be insurmountable mathematical difficulties. But Bohm showed how to overcome those difficulties, reviving de Broglie's idea of regarding a quantum object as a particle with an associated pilot wave that, in effect, tells it how to move.

In the pilot-wave picture, every quantum object is a real particle possessing definite attributes at all times. Associated with each such object is a pilot wave, which is real but is undetectable other than through its effects on the particle. This wave is termed the *quantum potential,* and serves the function of "reading" the environment and reporting its findings back to the particle. (It should be emphasized that this is a real wave and not to be confused with the quantum wave function, a purely mathematical gadget for making predictions.) The particle then acts in accordance with the information provided by its associated pilot wave. As a result, in the Quantum Potential Interpretation a quantum object is not composed of a single "thing," particle or wave, but is both. Notice how objective reality is restored in this picture, as there is no longer the ongoing schizophrenia between the object as wave or particle. At all times it is both, and at all times the particle side of the house possesses all the usual classic attributes. Bohm's genius was to show how this scheme could be made to work.

The most important theoretical development in quantum theory in the second half of this century is undoubtedly John Bell's discovery of a relationship that *any* system—quantum or otherwise—must obey. Let's look at a macroscopic system cooked up by D. Aerts to illustrate Bell's startling result.

Suppose we have two identical cubical containers V_1 and V_2 that are 20 centimeters on a side. Each container is assumed to be filled with 8 liters of water, and the two vessels are connected by a tube T that holds 16 liters of water. So the complete system $V_1 + V_2 + T$ holds 32 liters. The overall situation is shown schematically in Figure 7.3.

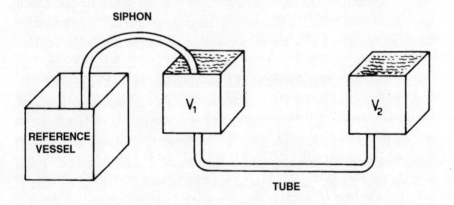

Figure 7.3. Aerts's example of a macroscopic situation that violates Bell's inequalities

There are two observables for each of the two vessels; call them O_1^a, $O_1^{a'}$, O_2^b, and $O_2^{b'}$. Each of these observables can take on the value $+1$ or -1, as follows.

• O_1^a: This measurement checks whether the volume of water that can be taken from vessel 1 is more than 10 liters. We carry out this observation by siphoning water from V_1 and placing it in a reference vessel. O_1^a has the value +1 if the water flowing from V_1 to the reference vessel exceeds 10 liters; otherwise $O_1^a = -1$.

• $O_1^{a'}$: This measurement checks whether the water is transparent. We take a glass of water from V_1 and hold it up to the light. If light gets through, $O_1^{a'} = +1$; otherwise, the value is -1.

• O_2^b: This observable checks whether or not the depth of water in vessel 2 is more than 15 centimeters. If it is, then $O_2^b = +1$; if not, then $O_2^b = -1$.

• $O_2^{b'}$: Here we check to see if the water in the second vessel is fit to drink. We take a spoonful of the water, drink it, and wait for five minutes. If we don't get sick, we set $O_2^{b'} = +1$; if we get ill within five minutes from drinking the water, $O_2^{b'} = -1$.

Now we want to look at the correlation functions between these observables. We denote by $P(a, b)$ the result of making measurement O_1^a followed by O_2^b, $P(a', b)$ the measurement $O_1^{a'}$ followed by O_2^b, and so forth. The order of experiments here is crucial, and we will assume that the measurement is made on vessel 1 first. The key idea is that the result of measurement O_2^b depends on the measurements made on vessel 1. If water is extracted, we will find that $O_1^a = +1$, because it is possible to extract 10 liters from vessel 1. But the level in vessel 2 will decrease from 20 centimeters to 10 centimeters, and we will then find $O_2^b = -1$. If instead the transparency of the water in V_1 is checked, the answer will be $O_1^{a'} = +1$, but the level in V_2 will have no noticeable change and, thus, $O_2^b = +1$. If we assume all the water is drinkable, we will of course have $O_2^{b'} = +1$, so the correlation functions have the values:

$$P(a,b) = O_1^a O_2^b = (+1)(-1) = -1,$$
$$P(a,b') = O_1^a O_2^{b'} = (+1)(+1) = +1,$$
$$P(a',b) = O_1^{a'} O_2^b = (+1)(+1) = +1,$$
$$P(a',b') = O_1^{a'} O_2^{b'} = (+1)(+1) = +1.$$

Bell's inequality says that if we add the magnitude of the difference $\|P(a, b) - P(a, b')\|$ to the magnitude of the sum $\|P(a', b) + P(a', b')\|$, the result cannot be larger than 2. Carrying out this calculation for the quantities just computed, this calculation yields the value 4—a violation of Bell's inequality. This result is a direct consequence of the nonlocal nature of the apparatus shown in Figure 7.3. Without the tube T connecting the two containers, there would not be any such violation. This tube plays the role of faster-than-light communication in the usual setup involving electrons and their spin rather than water tanks and their capacities.

The implication of these *Bell inequalities* is that any object that has definite properties at all times, as Einstein would have wished, must then participate in some type of faster-than-light signaling. Contrary to some interpretations that have been given of Bell's results, this does *not* mean that actual information can be transmitted in a superluminal fashion; so Einstein's speed limit for the universe is preserved. But it does mean that somehow particles that have interacted and then moved to different parts of the universe "know about each other"—instantaneously. Laboratory tests by Aspect and others in the 1980s confirmed Bell's results, which leaves physicists in the uncomfortable position of either having to abandon Einsteinian reality or accept faster-than-light signaling channels in the "deep structure" of the universe. We'll return to these matters in the Appeals phase of this trial.

In the 1980s, the experimental data available was completely consistent with *any* of the positions listed in Tables 7.1 and 7.2. So regardless of what view you held on either the Quantum Measurement or the Quantum Interpretation problems, the actual evidence would not refute you. The big developments in the last decade of quantum reality research have been observational. The hardware available for doing the kind of sensitive experiments needed to separate one of these positions from the others has finally become available, and hardly a month goes by nowadays without reports lending weight to one or another of these realities and discounting another. So most of the story in the Appeals will center on these developments, and how they enable us to weight our bets when it comes to sorting out how the world can *really* be like that.

THE APPEAL

Decoherence and the Cat's Kittens

Take a good look at the book you're now holding in your hand. Now put it down on the table and close your eyes for ten seconds or so and open them again. Chances are good that the book will still be there on the table. But was it there when you were not looking? If you answer, "Of course it was there; the book's behavior doesn't depend on whether I'm looking or not," then you haven't yet come to terms with quantum theory.

According to the dictates of the quantum mechanicians—classical-style, anyway—we have to remain outside a system in order to describe it, because in measuring the system we interact with it and thus change its quantum mechanical state in the process. In the account of the double-slit experiment given above, we have seen that for a quantum object like an electron, unlike a bullet or a baseball, we have to forsake the notion of a single path that it follows from the source to the detector. Rather, we must consider all possible paths that the electron might have followed and assess the probability for each of these various possibilities. It is the electron's wave function that measures these likelihoods. Let's see why the electron behaves differently when it's measured than when it is not by examining the double-slit experiment a bit more closely.

The setup is again shown in simplified form in Figure 7.4. Here any electrons that pass through the slits A and B can be detected at the screen, labeled XY. By blocking slit B, we would find that most of the electrons appear at the top half of the screen (curve 1); blocking slit A would yield curve 2. But if both slits are open, we might think what we'd see is simply the sum of these two curves (curve 3). But what we *do* see is not curve 3 but rather something like curve 4.

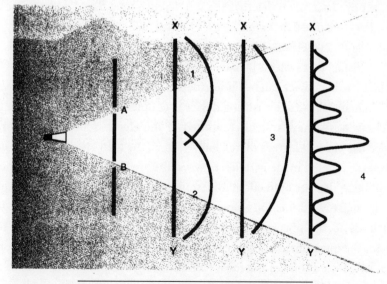

Figure 7.4. The simplified double-slit setup

This strange result means that some parts of the screen receive fewer electrons with both slits open than with one of the slits closed. Such an observation does not accord at all with the assumption that each electron must go through either slit A or slit B.

Traditional quantum theory says that there is a probability—call it x—for an electron passing through slit A, and another probability—y—for it to pass through slit B. The total probability for an electron to reach the detecting screen with both slits open is then $x + y$. But the rules of quantum theory say that the number of electrons that will reach any point of the screen is proportional to the square of this number, that is, $(x + y)^2$, which equals $x^2 + y^2 + 2xy$.

However, adding together the probability amplitudes telling us how many electrons reach the screen when just one of the slits is open gives only the number $x^2 + y^2$. So where does the extra $2xy$ come from? This term represents the interference between amplitudes x and y, and is a purely quantum-mechanical effect; it has no counterpart for baseballs and planets, just for quantum objects like electrons.

If we now make the experiment more sophisticated by tracking the path of each electron to determine which slit it passed through, we find that the wavy pattern of curve 4 disappears and we do indeed see curve 3. In other words, if we watch the electrons, each one goes by a "classical" path. Thus, the behavior of the system depends on whether we are observing it or not. The act of observation destroys the interference term $2xy$. From this we are forced to conclude that it must be the classical measuring apparatus doing the observing that is responsible.

The fact that systems behave differently when they are coupled to a classical measuring device can be understood in terms of a process called *decoherence*. When a classical measuring device is coupled to a quantum system like an electron, the new system composed of the electron and the measuring device has several degrees of freedom. That is, the total energy of the system can be distributed in many different ways. But we only ever use a small number of these degrees of freedom to describe the outcome of any measurement that we make. And it is the ignoring of large numbers of degrees of freedom that leads to the phenomenon called decoherence, which essentially suppresses the quantum interference between the different alternatives. To fix this idea, let's look at one of the most famous thought experiments in all of science, the celebrated Schrödinger's Cat.

Schrödinger's Cat

As soon as we artificially separate the system we're observing from the instrument that's actually doing the observing, the problems start. Shortly after developing the equation for the wave function that bears his name, Erwin Schrödinger devised the following thought experiment to illustrate the theoretical—and philosophical—difficulties that can arise. Here's the gist of his idea.

Suppose you put a cat inside a sealed box (A), which also contains a flask of poisonous gas (E) and a small chunk of unstable radioactive material, say a speck of radium (B). Inside the box is also a detector (C) that will register if the radium spontaneously emits a particle. If such an event occurs, a hammering device (D) breaks the flask, which releases the gas, and the cat (F) dies; otherwise, the cat lives on at least to the next instant. Figure 7.5 shows a cartoonist's version of the overall situation.

We set up this system at a given time and come back later and ask, "Is the cat dead or alive?" Is this the same as asking whether the radium emitted a particle or not? Standard quantum theory says the answer is that there was a certain probability (described by the Schrödinger equation) that it did and another probability that it didn't. So the quantum state of the system is a combination of yes and no. The same, then, goes for the cat, as well. It is both dead and alive, existing in a mixed state obtained by combining the two "obvious" states, dead and alive.

Figure 7.5. The Schrödinger's Cat thought experiment

Of course, nobody has ever seen a cat in such a state. When it comes to large systems like cats, we never see them in such weird combined states, even though they are perfectly legitimate in the quantum realm. But why don't we observe them?

The conventional answer is the one we gave earlier, namely, the "collapse" of the wave function when a measurement is actually made. By this view, when we open up the box and look inside to see the cat, at that very instant the wave function collapses into one of the two "pure" states, alive or dead. But until that measurement is made, the cat is in the combined state alive-dead.

This answer relies crucially on the existence of an external apparatus being used to make the measurement of the cat's state. And since we always assign a definite value to the measurement (not a probability distribution of possible values), such a measurement can only exist if the measuring apparatus obeys the classical laws of physics. This means that it must deal with the cat as a macroscopic—not quantum—object. This artificial division of a physical system into a quantum subject and a classical observer has always been the most unsatisfactory aspect of the conventional quantum-mechanical wisdom. It is what gave rise to von Neumann and Wigner's attribution of the role of human consciousness as the agent of the wave function's collapse. And it is also the source of Bohr's complementarity principle, by which an object can display wavelike or particlelike behavior—but not both at the same time.

To make progress on this conundrum, we have to reformulate quantum theory to eliminate the notion of an external observer. The way to do this is by invoking the physical process of decoherence, which makes macroscopic systems like cats behave the way we expect them to behave from common experience. In order to see how this process works, let's go back to the Schrödinger setup.

Generally, a system made of N particles—electrons, protons, billiard balls, or whatever—needs about $3N$ numbers to describe its quantum state. So for a cat weighing about one kilogram with 10^{26} atoms in its body, you would need around 3×10^{26} numbers to specify the state of the cat completely. These numbers would, essentially, characterize the wave function of all the particles of which the cat is composed. Such a description is purely quantum mechanical, and will tell you everything there is to know about the cat. But when you consider old Tabby sitting in Schrödinger's diabolical box, you describe her with far fewer parameters than this. In other words, there are *many* different quantum states of Tabby that will all be acceptable descriptions of the situation "Tabby sitting in the box."

This fact—that we can work with far less information than is needed to quantum-mechanically describe the cat—implies that the effect of unobserved (or, equivalently, unmeasured) parameters is to make the cat behave as though it is classical. So there is no fundamental difference between a classical system and a quantum-mechanical one. It's just that the system begins to behave more and more classically as we start ignoring large numbers of degrees of freedom. This suggests that if we could actually measure each and every one of the 10^{26} parameters specifying Tabby, then we'd find Tabby behaving just as much like a quantum object as an electron. In particular, she could then exist in a combined state in which she is both alive and dead.

It's important to note here that we have divided the set of parameters describing Tabby into two disjoint sets: those that we observe and those that we don't. And this division is quite arbitrary. In the jargon of the theoretical physicist, the unobserved quantities are termed the *environment,* and it is exactly the existence of these unobserved degrees of freedom in the description of the cat that washes out quantum interference and allows the system to behave classically. This is the process of decoherence. What begins as a system composed of a hopelessly tangled-up mess of quantum particles is untangled, or "decohered," by the simple process of ignoring all but a small number of the degrees of freedom in the system. What's more, decoherence is directly tied to the fact that classical systems are "large." The degree of classical behavior that a system displays is completely determined by the size of the system, relative, say, to the size of a completely quantum object like an electron. If the size is large, then the system will look and behave like a classical Newtonian system, a billiard ball or bullet. But if the system is small, then we can expect quantum interference to enter in an important way into its behavior.

This whole line of argument is superficially similar to another of Niels Bohr's famous principles, what's usually termed the Correspondence Principle. It states that the average behavior of a whole lot of identical quantum systems should mimic the behavior of a classical system. But decoherence addresses more the Principle of Complementarity, in which an object can be seen as either a wave or a particle, but not both. Now we see that this is just an extreme case of a more general principle. That is, that the object is *always* both, and that it is the degree of decoherence that determines whether we see it more as a classical particle or as a quantumlike wave.

Another crucial feature of decoherence is the role of observation. We have to make a choice as to which degrees of freedom to measure and which to ignore. In conventional quantum mechanics, all such observations assume a direct interaction between the system and the classical measuring device. But what if it were possible to actually make a measurement *without* interacting with the system at all? In that case, it should be possible to learn about the properties of the completely unobserved quantum object. And, in particular, this might give some insight into the puzzle of the reality of the wave function as a *material* object, not just a mathematical one. At first hearing, it seems impossible to conceive of how to measure something about a system without interacting with it in some way. But if there's one lesson to be learned from quantum theory, it's not to trust your intuition. Let's see why.

The Uncertainty of Uncertainty

On the streets of many big cities even nowadays you often see street hustlers trying valiantly to separate tourists from their cash by variations of the time-honored shell game. The hustler usually has three small cups and places a bean or a small marble underneath one of them. He then moves the cups around in a seemingly simple sleight-of-hand, with the gullible tourist asked to bet on which cup the marble is under. To almost no one's surprise, the tourist is soon fleeced and goes away a sadder but wiser observer of the street scene. Let's consider an interesting variant of the old shell game, as a way to illustrate how in quantum theory it may just be possible to measure something without interacting with it, at all.

Suppose we have two shells and a microminiature thermonuclear device that is hidden underneath one of the shells. Think of a country trying to hide its weapons from the prying eye of a spy satellite by moving them from one missile silo to another. The weapon is a very special one, however, since it will go off if even a single photon of light strikes it. The goal of the weapons inspector is to determine where the weapon is hidden but without exposing it to light or disturbing it in any other way. If the weapon goes off, the inspection has failed and the game is over.

At first hearing, this task seems hopeless. But a moment's thought shows that as long as the inspector is willing to be successful half the time, then an easy strategy is to look into the silo in which he hopes the weapon is *not* located. If he's right, then he knows the weapon is in the other silo, even though he hasn't actually seen it; if he's wrong, well . . . Winning with this strategy, of course, amounts to nothing more than a lucky guess.

Now let's simplify this game, but in such a way that a player limited to use of the laws and principles of classical physics has no chance to win. In this version, there is only one silo, and a random chance that the weapon is located in it. The inspector's goal is to say if the weapon is there, again without expos-ing it to light.

Suppose the weapon is in the silo. If the inspector doesn't look, then he gains no information. If he does look, then he knows the weapon is there—but only for an instant before it vaporizes both him and everything else in the vicinity. The inspector might try to dim the light so that there is very little chance of a photon hitting the weapon. But for the weapon to be visible, at least one photon has to strike it, leading to the same sad conclusion for the inspector. So what to do?

In 1993, two Israeli physicists, Avshalom Elitzur and Lev Vaidman, first offered a solution to this seemingly impossible conundrum. Their answer works, at most, only half the time. But as they say, half a loaf is better than no loaf at all. The method dreamt up by Elitzur and Vaidman makes crucial use of the fundamental nature of light, namely, that it can display both particlelike and wavelike behavior. Here's how it works.

Look, Ma, No Photons

As we've seen, the rules of quantum mechanics say that interference effects arise as soon as there is more than one possible way for a given outcome to occur, and the ways are indistinguishable by any means. In the double-slit experiment, an electron can reach the screen by two different paths (through slit A or slit B), and there is no way to tell which path it took. In this case, interference patterns arise on the detecting screen. But if one of the slits is closed, then the interference pattern disappears. So, put simply, in the absence of two (or more) indistinguishable paths, interference cannot occur.

As the starting point for their thought experiment, Elitzur and Vaidman consider an interferometer—a device with two mirrors and two beam splitters, as shown in Figure 7.6. Light entering the interferometer strikes a beam splitter, which sends the light along two paths, an upper and a lower one as shown in the figure. The paths meet at the second splitter, which then combines them and sends the light to one of two detectors, labeled D-Light and D-Dark in the figure. This is because one detector is positioned so that it detects only the equivalent of the bright fringes of an interference pattern (the D-Light detector), while the other detector sees only the dark fringes (the D-Dark detector). Here for the sake of definiteness, we assume that the apparatus is set up so that the photons always go to detector D-Light when the bomb is not present.

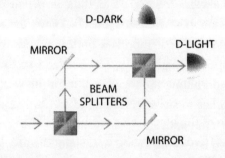

Figure 7.6. An interferometer for the Elitzur-Vaidman experiment

Now what happens if we place the hypersensitive nuclear weapon into one of these two paths, say, the upper one? If the first beam splitter acts randomly there is a 50 percent chance that the photon follows the upper path, in which case it hits the weapon, the bomb explodes, and everyone goes to their respective Valhallas. So in that situation the photon never gets to the second beam splitter.

But what if the photon takes the lower path, as shown in Figure 7.7? Now it doesn't hit the bomb. Moreover, interference no longer occurs at the second beam splitter, since the photon has only a single way to reach it. Consequently, the photon makes another random choice at the second splitter. It may then be reflected and hit detector D-Light, which gives no information, since it would have happened anyway even if the bomb had not been in the upper path. Or it may go to detector D-Dark. In that event, we know with certainty that the bomb was in one path of the interferometer, since if it was not, D-Dark could not have detected anything. So with this setup, we have somehow managed to make an interaction-free measurement; we have determined the presence of the bomb without interacting with it.

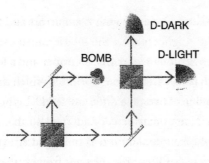

Figure 7.7. Path of a photon when the bomb is present

Clearly, this scheme only works some of the time. How often depends on randomness in the beam splitters. But when it works, it works completely. When the interferometer does not contain the bomb, the photon behaves as a wave. It can then reach both detectors simultaneously, leading to the characteristic interference pattern we associate with waves. But when the bomb is present, the photon behaves like a particle, following only one of the paths. So the mere presence of the bomb removes the possibility of interference, even though the photon need not have directly interacted with the bomb at all!

The Elitzur-Vaidman scheme just outlined is a thought experiment, similar in spirit to Schrödinger's fanciful thought experiment with the cat. But in some dazzling work by Anton Zeilinger and his colleagues at the University of Innsbruck in 1994–95, the thought experiment became a *laboratory* reality, thus verifying the predictions of Elitzur and Vaidman (but without benefit of a nuclear weapon, which was replaced by a mirror in the actual work of Zeilinger's group).

The experimental verification of the predictions of Elitzur and Vaidman led to another important step forward in understanding quantum reality. As we noted above, if the beam splitters send the photon in each direction with equal likelihood, then only half the measurements made are interaction-free. The question that arises is whether this is the best that can be done. None of the experimentalists could conjure up a way to improve the odds until Mark Kasevich of Stanford visited the Innsbruck group. During that stay, he suggested an approach based on the so-called quantum "Zeno effect" that makes it possible to detect the bomb in an interaction-free way *almost every time*. To see how, we need to know more about the Zeno effect.

The Quantum Zeno Effect

The Greek philosopher Zeno of Elea is best known for his "paradoxes" asserting that motion is just an illusion. For example, he argued that an arrow shot from point A will never reach a different point Z, since it must first reach the midpoint of the interval between A and Z (call it B). From there, it must reach the midpoint of the interval from B to Z, and so on forever. Thus, the arrow will never reach Z. Of course, Zeno refuted his own argument every time he went for a walk. But we can neglect this obvious counterargument. After all, Zeno *was* a philosopher.

By now, there are well-known responses to Zeno's "paradoxes" having to do with the infinite subdivisions of a finite—but continuous—quantity like an interval of distance or time. In quantum mechanics, however, there is a similar paradox that is very closely related in spirit to the argument put forth by Zeno.

Around 1977, physicists George Sudarshan and B. Misra of the University of Texas cooked up a theory showing that the decay of an unstable quantum particle like the nucleus of an atom of radium is suppressed by the simple act of looking at it. Remember that earlier we said that the conventional interpretation of quantum theory states that all possible properties of the atom are determined by its wave function. And what value you actually observe for a property like position or velocity is created by attaching a classical measuring device to the atom and "collapsing" the wave function. What we didn't speak of earlier was the state the atom is in after such a measurement. The standard reply is that it is in a new quantum state with its own wave function, which itself then evolves according to the dictates of the Schrödinger equation—until it is in turn collapsed by another measurement. The Sudarshan-Misra result says that this next collapse can be delayed by continuing to watch the atom. In other words, the more times the atom is observed, the greater is the suppression of the wave function collapse. And when it is observed continuously? Well, in that case the decay simply doesn't happen at all!

The implications of this result are quite amazing. For example, a radioactive nucleus that is constantly watched never decays, despite the fact that it is unstable. So perhaps the old adage "A watched pot never boils" makes some sense, after all. In the world of classical objects like pots of water, we know that watching has no effect on whether the water boils. But if it's quantum objects in the pot instead of zillions of molecules of water and hydrogen, things are different. Presumably, a single molecule of water would not boil if it were continually observed (if such a thing as a single molecule "boiling" makes sense at all).

For this quantum Zeno effect to occur, there is one condition that must be satisfied. For a short time interval after the unstable particle has been created, the probability of its decaying should increase with the square of its age. In practice, this condition is usually satisfied so it imposes no serious restriction on the applicability of the effect.

This short time interval is called the *Zeno time* for the particle, and if measurements are made within one Zeno time unit of each other, the wave function of the single particle never collapses and the decay is forever suppressed. The way this occurs is that initially the wave function is concentrated around the undecayed state. But as time goes on, the wave function gets smeared out into the decayed state. When an observation is made, the wave function "snaps back," or collapses, into the undecayed state again. So if observations are made on a short enough time scale, there is no possibility for the smearing out to occur and the particle is always seen in its undecayed state.

To test the theoretical predictions of Sudarshan and Misra, David Wineland at the National Institute of Standards illuminated ions of the atom beryllium with short pulses of light, each lasting only 2.4 milliseconds. These were the "measurements." He then measured the probability of the ions making a transition from a low-energy state to a high one. If the quantum Zeno effect was real, this probability should drop off exponentially as the number of measurements (pulses of light) per unit time increased. In Figure 7.8, the results of the experiments are compared with the theoretical predictions.

One might wonder whether these results offer an escape to the nuclear waste disposal problem. After all, if we could prevent radioactive nuclei from decaying, most of the difficulties associated with nuclear waste disposal would disappear, since then the waste wouldn't be dangerous. Maybe by just "looking" at the waste often enough, we could render it harmless. No such luck, I'm sorry to say, since the technology just doesn't exist for monitoring each and every atom on a time scale shorter than the vanishingly small Zeno time. So there's no help to be had from the quantum Zeno effect in neutralizing nuclear waste.

While the quantum Zeno effect is of no help in dealing with nuclear waste, it can help us dramatically improve the success rate in making interaction-free

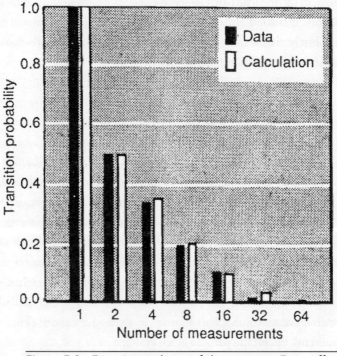

Figure 7.8. Experimental test of the quantum Zeno effect

measurements. Unfortunately, the procedure is a bit too complicated to describe here, so we have to refer the reader to the "To Dig Deeper" section of this book for details of how it's to be done. But the basic idea is to enable a photon to undergo a process that rotates its polarization. We then insert a screen to suppress the rotation of the photon's polarization. By implementing such rotation neutralization, in a setup similar to the Elitzur-Vaidman experiment, researchers at the Los Alamos National Laboratory have demonstrated that up to 70 percent of measurements can be interaction-free, a substantial improvement over the 50 percent seen in the original arrangement. And work currently under way hopes to increase this to a remarkable 85 percent.

Of what practical use are such interaction-free measurements? One intriguing possibility is for them to be used for imaging quantum objects by taking "photographs" of those objects without exposing them to light. Perhaps someday it will even be possible to X-ray patients without exposing them to X rays. If all this seems strange, it's well to recall that quantum mechanics operates in the realm of potentialities. So it is just because an interaction *could* have occurred that we can prevent one from occurring. Such measurements also shed considerable theoretical and philosophical light on the most publicly visible manifestation of traditional quantum mechanics, the celebrated Heisenberg Uncertainty Principle.

Heisenberg Demystified

By popular (mis)conception, the Heisenberg Uncertainty Principle asserts that the more accurately you want to measure one property of a particle, say its position, then the less accurately you can measure a complementary property like velocity. In the popular imagination, the reason for this uncertainty in measuring the particle's velocity is that the interaction by which the position is determined necessarily disturbs the particle and, hence, makes accurate measurements of the velocity impossible. The possibility of interaction-free measurements of the sort we just examined throws a dark cloud of doubt over this entire chain of reasoning, and should cause one to wonder about the so-called "inherent" uncertainty in quantum systems.

In the context of our benchmark double-slit experiment, conventional dogma says that an electron moving through the system is either a particle *or* a wave—but not both—depending on whether the slits are both open or one or the other is closed. And any attempt to ascertain both its particle and wave natures simultaneously runs afoul of the Uncertainty Principle. So what is this principle and why do physicists as illustrious as Feynman and Bohr appeal to it as the answer to the question, Where does the interference go in the double-slit experiment?

First of all, the Heisenberg formula for the uncertainty relation between complementary properties of a system is much older than the development of quantum mechanics. To see this, consider the uncertainty relation between the energy of a system, E, and the time interval over which this energy is measured, t. The Heisenberg uncertainty between these two properties can be written as

$$\Delta E \Delta t \geq h$$

where h is a fixed number called *Planck's constant*. What the formula means is that the error in the energy measurement times the error in the measurement of the time interval is always greater than or equal to the quantity h. Thus, the more accurately the energy is measured, the more uncertainty there is in the time interval over which the measurement takes place. It's very easy to derive this relation from principles of classical physics and optics. Here's how.

In 1900, Max Planck showed that the energy in a stream of electromagnetic radiation was related to its frequency, f, as

$$E = hf$$

Therefore,

$$\Delta E = h\Delta f$$

To obtain Heisenberg's formula, we need only appeal to the *principle of harmonic resolving power* from optics, which is even older than Planck's relation just shown. This principle says that if a wave of electromagnetic radiation of frequency f is chopped up by a time shutter into a stretch of duration Δt, then the width, Δf, of the spectral line will be

$$\Delta f \geq \frac{1}{\Delta t}$$

Substituting this relation into the one just given, we immediately arrive at the Heisenberg uncertainty relation between the two complementary properties, energy and time.

In this derivation, we interpret the symbol Δ to mean the spectrum of energies of the particles that make up the spectral line of the radiation. So the physical interpretation of the uncertainty relation is that if we vary the width of the time shutter (the quantity Δt), then we are bound to influence inversely the spectrum of energies of the particles making up the stream of radiation (ΔE). This interpretation, incidentally, shows clearly that we are talking about either a *population* of particles or sequences of many experiments with individual particles. In either case, the Heisenberg Uncertainty Principle is a *statistical* law, and says nothing about limitations on the precision of measurements of individual particles.

There is certainly no obvious reason why the Uncertainty Principle, which is basically a mathematical result, should be connected with something like wave-particle duality, which is a purely heuristic hypothesis. Perhaps it is due only to the magisterial authority of Niels Bohr, who was the first to relate the two via his Complementarity Principle. Such are the ways certain prejudices and misunderstandings get enshrined as dogma in the annals of science. Anyway, the situation is a whole lot more interesting than this, as the interaction-free measurements have finally produced some much needed clarity on the whole matter of waves versus particles by means of a theoretical gadget called the *quantum eraser.*

In the double-slit setup, the trick is to label each electron as it passes through a slit with a quantum-mechanical tag. If the tag were actually observed, then it would say unambiguously which slit the electron went through—but it's not observed. In contrast to the disruptive measurements of position discussed by Feynman and Bohr, such a tagging process is a microscopic operation, and as such is not subject to the Uncertainty Principle. Theory predicts—and experiments confirm—that with such a tagging system in place no interference pattern develops. What's responsible for wiping out the interference, then, is not knowledge of what slit the electron went through, but the mere *possibility* of finding out. Where does the eraser come into play?

The eraser is installed downstream, between the slits and the screen. Its function is to tear the tags off the electrons *after* they have passed through the slits, but *before* they reach the screen. Once that is done, miraculously the interference pattern reappears. But has the Uncertainty Principle really been circumvented?

If Feynman were still alive, he might interpret this result by arguing that a tagging device, *subject to the Uncertainty Principle,* determines which slit an electron passes through. This establishes the electron as a particle and scrambles its path. The eraser then undoes the disturbance introduced by the tagging and restores the electron to the wavy state it possessed before it was tagged. But such an explanation seems cumbersome and unconvincing. It is more reasonable simply to assume that the Uncertainty Principle has not entered into the situation at all, suggesting that electrons are both particles and waves at the same time.

Further theoretical developments of this quantum eraser idea by the German physicist Berthold-Georg Englert support this conclusion. Englert declared that waviness and particulateness can be measured, provided you define them in the right way. His proposal is that instead of waviness, physicists should speak about *fringe visibility,* which measures the quality of the interference pattern. When fringe visibility is high, the stripes are clear, and their cause is a wave; when it is zero, the stripes are totally absent, and the pattern could have been created by bullets. In place of the black-and-white property of particulateness, Englert suggested a variable called *predictability* of which-way information. This is simply a measure of how sure you are that a given particle went through a chosen slit.

Surprisingly, analysis of these two variables—fringe visibility and predictability—showed that they are *not* related by an uncertainty principle. In fact, they are not even separate quantities, like position and velocity, but different manifestations of the same variable. At one extreme of that variable, the fringes are sharp, which-way information is absent, and what is going through the slits is a wave. At the other extreme, the fringes are absent, the electrons can be assigned to definite slits, and so they are particles. In between resides an entire spectrum of descriptions, such as "2 percent wave and 98 percent particle," or "half wave, half particle."

So particulateness and waviness turn out to be degrees of the same underlying property of an object. The physicist Hans Christian von Bayer likens it to a situation in which physicists studying sound waves had identified two separate properties—loudness and softness—and concentrated only on the extremes. In that case, they might have come up with a complementarity principle saying, "A loud sound lacks softness, and a soft sound lacks loudness." An analogous uncertainty principle may have been stated, "If you measure a loud sound perfectly, then you can't know anything about a soft sound." Of course, there is really only a single variable, volume, which can be changed continuously from one extreme to the other.

The lesson from all this is that an object like an electron or a baseball is neither a wave nor a particle, but both and neither. They have a quantum-mechanical character that is observed as waviness at one end of a spectrum and as particulateness at the other end. Presently, we have to try to develop our intuition further to understand what it means to be in-between. Now that we are starting to develop mathematical theories for this spectrum of possibilities, we can confidently expect to see the traditional arguments of Bohr, Heisenberg, Feynman, and others somehow folded into a single theory that will show that what you see depends entirely on how you look.

The quantum eraser and interaction-free measurements certainly seem to be doing away with traditional quantum uncertainty, Heisenberg-style. But there is another possible source of uncertainty that must also be considered, quantum chaos. The Schrödinger equation describing the evolution of the wave function of an object is completely untouched by the foregoing discussion. As the Schrödinger equation is a classical, deterministic dynamical system, one might wonder to what degree the phenomenon of chaos, which imposes limitations on our ability to successfully track the trajectories of such systems, might pose inherent limits of its own on what we can know about a quantum system. Let's briefly look at this possibility.

Quantum Chaos

Probably the most important development in applied mathematics in the last fifty years has been the discovery (or rediscovery, actually) of what has come to be called "chaotic" phenomena. The canonical metaphor for all such processes is the famous butterfly in Brazil, which by flapping its wings can lead to hurricanes in Florida. The essential ingredient of all such systems, then, is that a very small, minuscule even, change in the initial state of the system can be magnified into a major change in the system's behavior at a later time; in essence, the system is pathologically unstable.

The instability in chaotic systems leads to the surprising result that a system whose rule of behavior is totally deterministic, containing not a whit of randomness, can still display behavior that is indistinguishable from the random toss of a coin or the roll of a die. Thus, unless the initial state of the system is known *perfectly,* and the rule of change of state is executed *without the slightest error,* an unavoidable uncertainty is introduced into any prediction of where the system's state will be found at any given time in the future. Again let me emphasize that this uncertainty is due not to any type of intrinsic randomness in the system, but only to our inability to measure things to an infinite level of precision. For example, almost every real number is irrational, which means that it cannot be written as the ratio of two whole numbers. Put another way, the decimal expression for such a number will go on forever. So if the initial state of a system is characterized by some real number, the chances are overwhelmingly high that we will never be able to specify that number *exactly*. The best we can do is give a close approximation. But this error, however small, will soon be magnified at some future time by the chaotic nature of the system's dynamics into a major departure of the state from what it would have been if the error had not been made.

So here we have a fundamental source of unpredictability in nature, the *butterfly effect*. Heisenberg uncertainty is another such limitation on our ability to predict with precision. Is there any connection between the two?

While both chaos and the Uncertainty Principle impose fundamental limits on what can be known about the world, the unpredictabilities in quantum theory and chaos are very different in kind. The problem is that in quantum mechanics small perturbations in the initial state of the wave function generally only lead to small perturbations in subsequent states. So it's difficult to see how there can be any chaos in quantum systems. But as always with quantum theory, things are never simple and straightforward.

The British physicist Michael Berry argues that the difference between the chaos that we see in classical systems like balls moving about on a billiard table and quantum systems like an electron is that quantum uncertainty imposes a fundamental limit on the sharpness of the dynamics. We saw earlier that the Heisenberg uncertainty is quantified by a quantity, h, known as Planck's constant. Berry says, "In classical mechanics, objects can move along infinitely many trajectories. This makes it easy to set up complicated dynamics in which an object will never retrace its path." And this is just the sort of behavior that leads to chaos. But in quantum mechanics, Planck's constant blurs out the fine detail, which smears out the chaos.

So what happens if you scale down a classical system to atomic dimensions? Does the chaos smoothly fade away into a kind of quantum regularity? Or does the chaos remain, even though we can't observe it in the quantum system? And why are macroscopic systems like billiard balls chaotic given that everything they are composed of is atomic objects and thus quantum in nature?

One answer to some of these questions was provided by physicist Martin Gutzwiler of the IBM Research Laboratories. He produced a key formula showing how classical chaos might relate to quantum chaos. The essence of Gutzwiler's result is that regularities in the quantum behavior are related to a very limited range of classical trajectories, namely, those that are periodic. For example, if you played billiards on an oval-shaped table and hit a ball on exactly the right path, you could get it to retrace its path after only a few bounces off the sides of the table. But because this classical system is chaotic, these paths are very unstable; a minuscule error in striking the ball will send it off on a completely different course after only a couple of ricochets. So classically you would not expect to see these periodic orbits—and you don't! But a game of quantum billiards using electrons instead of ivory balls would be a very different thing. The reason is that the uncertainty in quantum mechanics "fuzzes" out the trajectory of the ball, so that tiny errors don't mount up as they do in the classical realm. So the periodic orbits are reinforced in some way we don't yet completely understand, which results in the periodic orbits being emphasized rather than suppressed.

The situation is actually a bit more complicated than this, however, as quantum mechanics really only fuzzes over the chaos for a certain amount of time, called the *quantum break time* of the system. This time turns out to be the crucial factor distinguishing between quantum and classical predictability of chaotic systems. Before the break time, quantum systems can mimic classical systems by looking essentially random. But after the break time, the system becomes periodic, retracing its path over and over again.

If this explanation involving the quantum break time is correct, how can we rationalize the existence of chaos in classical systems? The resolution involves returning to the fact that classical systems are composed of atoms, which of course are quantum objects. As it turns out, classical systems are in fact behaving like quantum systems. The only difference is that the quantum break time for macroscopic systems like billiard balls is enormously longer than the age of the universe. So if we could study a classical system for a period longer than its quantum break time, we would see it settle into a behavior that is not really chaotic at all. It would be quasi-periodic instead.

The end result is that classical and quantum realities can be reconciled, after all. The classical world is simply embedded within a larger quantum reality, one that manifests itself only in the realm of the very small and the very quick. In short, anything classical mechanics can do, quantum mechanics can do better.

The leitmotiv of our story so far has been that objects—classical and quantum—are neither pure waves nor pure particles but some combination of both. Of all the quantum realities, the one that most closely holds to this picture is the pilot-wave theory first presented by Louis de Broglie in the 1930s, and later developed by David Bohm. It's worth reexamining this reality in light of the past decade's worth of developments in quantum reality research.

Wavicles

As we have seen, the standard quantum theory party line holds that there are two entirely different laws governing the behavior of an object. One law, the Schrödinger wave equation, talks about how the object's behavior unfolds *when it is not being observed.* This law is entirely deterministic. The other law holds that what you will see on your measuring device if you actually observe the object differs from the properties of the object when not being observed. This law governs the so-called "collapse" of the object's wave function when an observation is made, and is inherently probabilistic. The Measurement Problem of quantum theory is essentially involved with the way in which we distinguish those conditions under which the first type of law applies from those conditions when we must use the second kind. All classical quantum theory has to offer in this regard is some vague notions about the distinction between the observer and the system being observed, that is, between subject and object.

Of all the prospective solutions to the Measurement Problem that have been offered, the only one that is completely deterministic and that denies the seeming absurdity of cats that are both dead and alive is one proposed by the late David Bohm, the so-called pilot-wave theory. The details of this proposal were covered in *Paradigms Lost,* so let me just briefly summarize them here.

Bohm's first daring assumption is to regard the wave function of the quantum object as being not just a mathematical artifice, which we massage in order to get predictions about the quantum object's properties. Rather, Bohm's theory demands that we grant actual physical reality to this wave function. He thought of it as a kind of field, what he called the *quantum potential.* This field is much like the gravitational or electromagnetic fields familiar from classical physics. The theory also regards a quantum object as a particle that has definite location in space at each moment. The quantum potential then pushes its associated particle around, to guide it along its proper course.

The mathematical formulation of the pilot-wave theory, then, involves not only the familiar Schrödinger equation for the wave function, but also other laws dictating how the wave function moves its particle from one place to another. All those laws are completely deterministic. Thus, the positions of the particles at a given moment can, in principle, be calculated with certainty from their positions at any earlier time—just as in classical mechanics. What we have then is a view of the quantum object as a kind of "wavicle," consisting of both a *real* particle and a *real* wave that serves to guide, or "pilot," the particle.

Bohm's theory interprets the double-slit experiment by asserting that an electron definitely goes through one of the two slits, A or B. Period. Which slit it will pass through is fully determined by the electron's initial position and the initial state of its wave function. But regardless of which route the electron takes *as a particle,* its associated wave function will split up and take both. So, for example, if the electron goes through slit A, it will nevertheless be reunited with the part of its wave function that went through slit B. Even more strangely, once the two parts of the wave function are reunited, the part that went through slit B can "inform" the electron of what things were like along that path that the electron itself did not take. Moreover, the theory entails that this "empty" part of the wave function that went through slit B is completely undetectable; only the part that went through slit A along with the electron can have any effect on the motions of other particles.

The pros and cons of this pilot-wave theory are outlined in *Paradigms Lost.* What's important for us here is its unbridled assertion that the wave function has a physical reality every bit as real as the particle itself. A few years ago, Yakir Aharonov, a former student of Bohm's, and Jeeva Anandan proposed an experiment to test this hypothesis. Here's how it goes.

The Quantum Wave Function—Mathematical Fiction or Physical Reality?

The celebrated Princeton physicist John Archibald Wheeler once remarked, "The quantum wave has the same relationship to a particle as a weather report has to snow," as a joking reminder of the mathematical nature of the wave function and the physical nature of a particle. And we all know that you can't get wet by reading the weather forecast in the newspaper. Nevertheless, Aharonov and Anandan constructed the plan for an experiment that if it works would indeed show that the quantum wave is just as real as the particle it represents.

Their scheme relies on sending a neutron zipping through a spatially varying magnetic field toward a screen that displays the point of impact. The neutron acts as a very weak magnet by virtue of its spin and thus feels a slight force proportional to the rate of variation of the field. The direction of the neutron's spin determines the direction of the force, so an *up* neutron is pulled upward by the force, a *down* neutron is pulled down below the center of the screen. But because the neutron's quantum wave function combines both up and down states, the wave ordinarily splits along two paths creating both high and low spots on the screen. So the measuring apparatus gets entangled with the neutron, and in the conventional view it collapses only when the screen is observed, showing either a high or a low spot.

Aharonov and Anandan have proposed a sort of "protected" version of this experiment, in which the varying magnetic field is kept extremely weak, while a second, nonvarying field is added. This second field doesn't deflect the neutron because it is not varying. But it does influence the wave function because its intensity is great enough that it swamps the effect of the weak magnetic field. So it can't split the wave into the two parts, up and down. As a result, the wave strikes the screen as a single spot. Thus, in this setup the measuring device doesn't get tangled up with the wave function, and the screen can be observed without collapsing the wave.

Even though the weak field is unable to split the neutron's wave, it still exerts a tug on it, so that the spot formed on the screen is off center. Just how far off center the spot ends up reflects all the various potential states represented by the wave. So the location of the spot serves as a surrogate measure of the uncollapsed wave itself. As of this writing, experimentalists have not yet taken up the challenge of how to perform a measurement on a single neutron or atom. But as Anandan says, "Experimentalists always seem to have a way of surprising me." When they do, the reality of the quantum wave function may well turn out to be the surprise package he's waiting for.

The reality of the wave function, should it finally be demonstrated experimentally, would force a major revision of the way theorists think about quantum mechanics. And one of the most central revisions would be in their view of what we earlier termed the Quantum Interpretation Problem. This refers to the question of whether it makes sense to think that particles like electrons have a well-defined position and velocity at all times—even when they are not being measured. The famous result of John Bell, relating this question to the speed limit of the universe as postulated by Einstein, has seen major developments and extensions in the past decade aimed at sharpening our understanding of what the question actually means. Here we'll examine a few of the highlights.

To Be or Not to Be—That Is Not the Question

One of the most cherished principles of Aristotelian logic is the law of the excluded middle; things either are or they're not. Cats are either alive or dead; electrons go through either the left slit or the right one; photons either spin up or they spin down. But in quantum mechanics "or" is replaced by "and." And thus goes the law of the excluded middle. Or does it?

In the early days of quantum theory, Einstein felt that the replacement of "or" by "and" reflected an inherent incompleteness in the description of nature on offer from quantum mechanics. His feeling was that while the theory did an exemplary job of predicting the outcome of experiments, it failed in providing a full and complete account of the properties of material objects. For example, Einstein rejected the notion that an electron goes through both slits of the double-slit apparatus when the slits are both open. It goes through one or the other, he said, so if quantum theory says otherwise there must be something wrong with the theory. Einstein's argument was that there must be "hidden variables," not part of the existing theory, whose values if known would give a full account of the behavior of an electron—and every other material object. In short, the quantum theory as it stood (and still stands) is incomplete.

EPR, Superposition, and Entanglement

To illustrate his discomfort with the existing state of affairs in quantum theory, in 1935 Einstein, along with Boris Podolsky and Nathan Rosen, two colleagues at the Institute for Advanced Study in Princeton, constructed a thought experiment that brought into sharp relief the failings of the quantum theory. The gist of this Einstein-Podolsky-Rosen (EPR) experiment (as later simplified by David Bohm) goes as follows.

Suppose we have a single unstable elementary particle, a pi meson, that decays into an electron and a positron. These two particles fly away from each other in opposite directions (so that linear momentum is conserved), toward observers Alice and Bob who measure the spin of their respective particle. Suppose that Alice measures the electron's spin and finds it to be spinning *up*, while Bob measures the positron's spin. Since the pi meson had no spin at all,

and the total spin before and after the decay must be conserved, Bob will necessarily measure the positron's spin to be *down*. According to quantum
mechanics, neither Alice nor Bob can know beforehand what they will measure; a priori, up or down is equally likely for both. Yet Alice knows that whatever measurement she makes, Bob's measurement must be the opposite. This
mysterious connection between the spin of Alice's and Bob's particles is called
entanglement in the quantum theorist's jargon. It's illustrated in Figure 7.9.

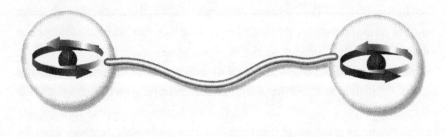

Figure 7.9. Quantum entanglement of a pair of particles

In the EPR paper, the authors made an argument that's now called *realism*. Its essence is that at the instant Alice measures the electron's spin as up,
the spin of the positron becomes an element of physical reality; it has an
objective, real existence *whether or not Bob decides to measure it.* But Alice's
measurement cannot influence any of the properties of Bob's positron. Her
electron can't send an instantaneous signal to the positron, telling it which
value of spin to adopt. So if Bob and Alice make their measurements simultaneously, the results of one cannot affect the other. EPR insist that the two systems—the electron and the positron—are separate. This separation is called
locality, and the combined principles of realism and locality are called *local
realism.* In short, local realism says two things: The two objects have definite
spin values *at all times,* regardless of whether they are being measured (realism); measurement of the spin of one of the particles cannot instantaneously
affect the measurement of the other (locality).

Classical quantum theory says otherwise. It claims that no description
of the particle's quantum-mechanical characteristics can specify the spin
value of a particle before it is measured. According to this view, the electron's
spin is in a *superposition* of the two states, *up* and *down,* and which is the case
is determined only at the time the electron is actually measured. This situa-

tion is shown schematically in Figure 7.10, where the electron is viewed as a tiny magnet whose field can point up or down. As a result of this ambiguity in the electron's spin, Einstein concluded that quantum mechanics must provide an incomplete description of the situation, since the EPR experiment says that elements of physical reality could exist when quantum theory says they can't.

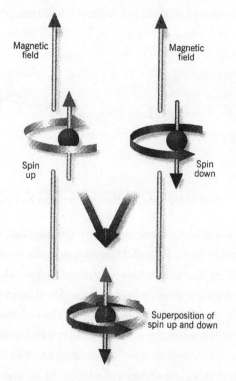

Figure 7.10. Superposition of spin states for an electron

Shortly after publication of the EPR paper, Niels Bohr objected to the argument saying that the only "real" things are those that we can measure. According to Bohr, the strangeness of quantum mechanics is simply a reflection of the fact that making some measurements precludes making others.

The next major step toward understanding the EPR experiment was in 1964, when Irish physicist John Bell published a paper placing an upper limit on how well correlated Alice's and Bob's measurements could be if they measured their particle's spins along different directions. Bell derived his famous *Bell inequalities* assuming the principle of local realism. So what we have is that local realism implies satisfaction of Bell's inequalities. But quantum theory, classical-style, predicts that in some cases the inequalities will be violated. And, in fact, experiments performed by John Clauser, Alain Aspect, and others in the early 1980s showed that indeed the Bell inequalities were violated. Hence, good-bye to local realism.

It's a good thing Einstein had died by the time these results were discovered, since he would have hated their implication. Either reality would have to go, meaning that particles do *not* have well-defined properties at all times, or one would have to accept that the measurement of a property of one particle could instanteously affect properties of another, possibly far-distant, particle. This latter would seem to be a violation of Einstein's own edict on the velocity of light as an upper limit to the speed with which any signal can be transmitted from one location to another. What a dilemma!

Dutch-Door Physics

To further seal the coffin on local realism, in 1993 Lucien Hardy of the University of Durham in the U.K. and Thomas Jordan of the University of Minnesota at Duluth in the U.S.A. suggested another thought experiment to demonstrate the conflict between quantum mechanics and local realism—and without making any use of the Bell inequalities. Here's how it goes.

Alice and Bob measure the spins of their particle along two different directions, A1 or A2 for Alice, B1 or B2 for Bob. As before, the spins in these directions can be only *up* or *down*. If these directions are selected appropriately, four propositions must be true:

1. If Alice measures *up* along direction A1, then Bob will measure *up* along B2.

2. If Bob measures *up* along direction B1, then Alice will measure *up* along A2.

3. Sometimes, Alice and Bob will measure *up* along A1 and B1, respectively.

4. Alice and Bob will never both measure *up* along A2 and B2, respectively.

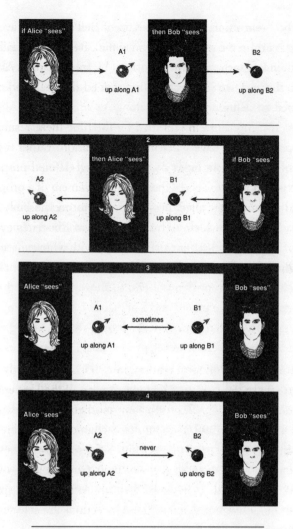

Figure 7.11. The Hardy-Jordan experiment

The overall situation described by these propositions is shown pictorially in Figure 7.11.

If we assume local realism, these propositions generate a contradiction, since they can't all be true simultaneously. Here's why. Proposition 1 says that when Alice measures spin up along A1, Bob will see spin up along B2. Thus, since Bob's value of the spin can be predicted with certainty, his particle's spin along B2 must exist whether or not he decides to measure it. Moreover, local realism demands that this be true independently of Alice's choice of direction, A1 or A2, and thus must have existed from the moment Bob's particle was created. Similarly, Proposition 2 says that Alice's particle must have possessed spin up along A2 from the moment it was created. By Proposition 3, Alice and Bob may sometimes measure spin up along A1 and B1, respectively. When this happens, the first two propositions say that both Alice's and Bob's particles must have had definite spin values along A2 and B2. So if Alice and Bob had chosen to measure along A2 and B2 instead, they would have found the spins to be up along those axes. This, in turn, violates Proposition 4. Thus, the assumption of local realism leads to a flat-out logical contradiction among these four propositions.

An everyday analogy involving Dutch doors reveals how the Hardy-Jordan argument conflicts with local realism even more vividly. Imagine a pair of Dutch doors that are latched, so that the upper and lower sections move together. In this analogy, open and closed doors will represent spin up and spin down, respectively, and Propositions 1 and 2, above, will be used to label and open the doors, as shown in parts 1 and 2 of Figure 7.12. According to the third Hardy-Jordan proposition, the two bottom doors will be open together some of the time (part 3 of the figure). However, proposition 4, which states that the two top doors can never open together, illustrates the contradiction, since no latched Dutch doors can meet all four propositions at once. If both bottom doors are open, both top doors will have to be open as well, which violates Proposition 4.

Hardy and Jordan found quantum-mechanical systems that *do* satisfy all four propositions, simply because quantum systems can violate local realism. In terms of the Dutch-door analogy, quantum mechanics only determines the state of any two doors that are actually observed, and says nothing about the two doors that are unobserved. This would be Bohr's explanation. The paradox arises only if one insists that all four doors must exist regardless of which ones are actually being observed. Experiments carried out by David Branning and his colleagues at the University of Rochester have confirmed the Hardy-Jordan approach, but not without a loophole or two. Nevertheless, the evidence is mounting that the world does indeed violate local realism.

Quantum Spookiness

In March 1947, Einstein wrote to the German physicist Max Born, saying, "I cannot seriously believe in [the quantum theory] because it cannot be recon-

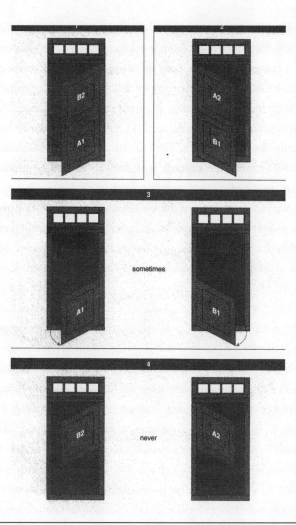

Figure 7.12. A Dutch-door analogy for the Hardy-Jordan experiment

ciled with the idea that physics should represent a reality in time and space, free from spooky actions at a distance." Ironically, in 1997 researchers from the University of Geneva demonstrated the reality of quantum spookiness less than one hundred miles from Bern, where Einstein did some of his most famous work.

The Swiss team headed by Nicolas Gisin demonstrated quantum action-at-a-distance on a large scale, by creating pairs of entangled photons and sending them through optical-fiber lines to the two small villages of Belle-vue and Bernex, which are nearly seven miles apart, where the photons were analyzed and counted. When the photon counts were relayed to Geneva via a second optical-fiber link and compared, they turned out to be correlated. Thus, each photon in a pair knows what its distant partner is doing—and does the same thing. The implication of this is that certain properties of the photon twins are not defined at the moment the pair is created. The photons acquire a particular state only when a measurement is made on one of the pair, which instantly determines the state of the other photon. This result is pretty definitive proof that the entanglement of the photons at the moment of their creation does not fall off with distance. So events in the far corners of the universe might well influence events here on Earth. This fact, in turn, leads to a plethora of possibilities for teleportation, cryptography, and computation. Let's close this discussion of quantum reality by examining a few of these possibilities.

Teleportation, Q-Bits, and Computation

One of the most enduring technological marvels of the *Star Trek* movies and television series is the "transporter," that mythical device by which Scotty and his merry band of engineers manage to beam Captain Kirk from the bridge of the *Enterprise* to the surface of alien worlds. While such instantaneous transmission of matter is still but a dream in the world of macroscopic objects like us, teleportation is now a reality in the quirky world of quantum mechanics.

In 1993, IBM physicist Charles Bennett, along with several colleagues, noted that a physical object is equivalent to the information needed to construct it. Therefore, they reasoned, the object can be transported by transmitting the information along any conventional channel of telecommunication, the receiver then using the information to reconstruct the object. In the classical realm, there is nothing earthshaking about this idea. In fact, it is exactly what happens when you send a fax. But the principle of the fax machine can only work with macroscopic objects; as soon as the scale of faithful reproduction of detail begins to shrink, eventually you run afoul of the Heisenberg Uncertainty Principle of quantum mechanics, the consequence being that the object's original state is bound to be destroyed by the process of scanning it for transmission.

Bennett et al. circumvent this barrier by transmitting quantum information using the entanglement of two quantum objects. The object to be transmitted, C, is entangled with another object, A, by a quantum measurement of joint properties of the two particles. Normally, this measurement act would destroy the full quantum information about C. But if A is already entangled with a third object, B, the effect of the measurement of A and C is to transfer the state of C to B. This transference is not perfect, but is the result of a transformation that depends on the result of the measurement. This transformation can then be undone by someone who knows the results of the measurement, leaving B in the same state that C started in. B has then become the transported version of C. The overall situation is shown in Figure 7.13.

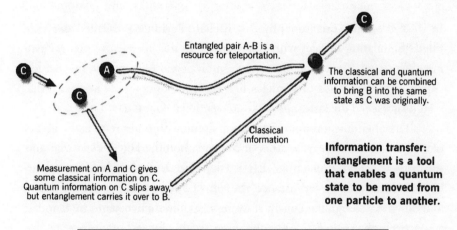

Entangled pair A-B is a resource for teleportation.

The classical and quantum information can be combined to bring B into the same state as C was originally.

Classical information

Measurement on A and C gives some classical information on C. Quantum information on C slips away, but entanglement carries it over to B.

Information transfer: entanglement is a tool that enables a quantum state to be moved from one particle to another.

Figure 7.13. Bennett's scheme for quantum teleportation

So suppose Alice has a message photon whose spin state (up or down) she wants to transmit to Bob. She makes a combined measurement on one member of an entangled pair of photons, together with her message photon. This measurement doesn't tell her the state of her message photon. But it does entangle her two photons in such a way that they end up with opposite states. As a result of her measurement of this entangled pair, Bob's photon becomes instantaneously primed so that when Alice tells Bob what to do to his photon, he will find that it is identical to the original message photon.

Interestingly, this process destroys the original state of Alice's message photon. But that state lives on in Bob's photon, irrespective of the distance between Alice and Bob. So there is no transport of actual photons between them, only a flow of quantum information.

In late 1997, Anton Zeilinger and his colleagues at the University of Innsbruck in Austria performed experiments that demonstrated the validity of Bennett's idea. But sadly for *Star Trek* fans, quantum teleportation cannot be scaled up to move Captain Kirk from one place to another. It's more like teleporting "states" of Kirk around the universe, rather than moving Kirk himself. But teleportation of this type has another, perhaps even more important application—moving data around inside a quantum computer.

Quantum Computation

In 1982, the famed quantum physicist Richard Feynman published an article titled "Simulating Physics with Computers." In this speculative piece, Feynman suggested harnessing the weird ambivalence of quantum-mechanical superposition of states of particles like electrons to compute at a pace that would far exceed the fastest possible conventional computer.

The phenomenon upon which a quantum computer rests is the ability of a quantum system, say an atom or a single photon, to be in more than one quantum state at the same time, that is, a superposition of states as shown earlier in Figure 7.10. So, for instance, the spin of a photon can be in both the up and down states simultaneously. If we represent these two states by 0 and 1, respectively, then calculations on the superposition act on both values at once. Thus, a quantum computer containing n photons or atoms in superposed states could do a calculation on 2^n numbers simultaneously—a degree of parallelism that is inconceivable for everyday, classical computers.

There are a couple of drawbacks to this rosy picture, however. The first is that the process of quantum-mechanical measurement limits (via the Heisenberg Uncertainty Principle) the amount of information that can be extracted from a quantum computer. The second is that quantum superpositions are delicate, fragile things; any contact with the environment sets off the process of *decoherence,* which collapses the superposition to just one of the many possible states the quantum object can occupy. This collapse, of course, eliminates entirely the advantage of using a quantum, rather than classical, computer.

A good question to ask is, Why bother trying to build such a gadget as a quantum computer? Until rather recently, there was no convincing evidence to support the idea that a quantum machine could compute anything that could not be done equally well with a classical computer. But in 1992, Peter Shor of AT&T Bell Laboratories published a quantum-mechanical algorithm for factoring large numbers into their prime components that showed the superiority of using quantum computers over their classical counterparts. Factoring large numbers is a task of not only theoretical, but practical, importance in cryptography, which resulted in Shor's work getting a lot of attention. In all known classical factoring algorithms, the amount of time needed to find the prime factors of a number grows as an exponential function of the size of the number (for those with a bent for figures, this factor is approximately $\exp(L^{1/3})$, where L is the length of the number in question). Shor's quantum algorithm requires a time proportional to L^2, a polynomial growth in L. This is a dramatic reduction over the classical methods when L becomes large. So dramatic, in fact, that it transforms factoring from a computationally "hard" to a computationally "easy" problem.

The question surrounding the feasibility of Shor's method in practice (leaving aside the not entirely simple question of actually constructing a quantum machine) is whether the exponentially fast collapse of the superposition of states when exposed to the environment would swamp the exponential increase in computing speed for the factorization scheme. Careful estimates by Shor, Woijcech Zurek, and others have shown that quantum computation can still be useful in factoring as long as a sufficiently low decoherence rate can be maintained. Basically, these results rely on Shor's discovery of quantum analogs of the error-correcting codes used in conventional computers. The standard schemes for error correction cannot be used in the quantum case because they involve actually reading the state of the computer as the computation unfolds and creating redundant states to get any erroneous calculation back on track. The difficulty in the quantum situation is that information disappears in a quantum system as soon as you look at it.

The trick to creating an error-correcting scheme for the quantum computer is to encode the state of a single quantum bit as a combination of states in a multiple-bit system. In other words, Shor spreads the information in a single bit across nine such bits. In this way, the original quantum information is preserved even if an error occurs in one of these nine bits. Seth Lloyd of MIT, one of the pioneers in the quantum-computation game, states that the existence of such error-correcting codes came as "quite a surprise. Before

Shor came up with this idea, nobody thought it was possible." One problem, however, noted by Lloyd is that error correction, being a computation all its own, runs the same risk of making mistakes as in the original computation. At the moment, nobody has any good idea of how to address this "second-order" difficulty in quantum computation.

Unfortunately, there is no room here to enter into the technical details of various approaches people have presented for actually constructing a quantum computer. But there appear to be no physical or logical barriers to be over-come—just massive engineering and technological ones. So while it may be some time before we find ourselves with a quantum machine on our desks, happily buzzing away factoring numbers several hundred digits in length, the possibility for such machines is clearly evident. And if the history of technol-ogy teaches us anything, it's that once something is possible, it becomes almost mandatory. As Anton Zeilinger says, "Information is deeper than reality."

Fascinating as these developments in quantum information are, they don't really bear directly on our basic quest for uncovering the "true" nature of quantum objects. So let's return now to the problem of quantum reality, and take a harder look at something new.

Consistent Histories

In his book *Philosophical Investigations,* the Austrian philosopher Ludwig Wittgenstein remarks, "The table I see is not made of electrons." This senti-ment agrees with our everyday macroscopic observations about tables, but seems at odds with the worldview of quantum mechanics. The conflict, it appears, lies in the impossibility of identifying in quantum logic a classically meaningful property like the color and hardness of the material composing a table with a set of elementary properties referring to the microscopic objects making up the table (e.g., the electrons in the table's atoms).

Recently, a number of researchers, including Robert Griffiths, Wojciech Zurek, Murray Gell-Mann, James Hartle, and Roland Omnés, have proposed interpretations of quantum mechanics that do not depend on having an exter-nal observer to describe the properties of a system. Since the schemes pro-posed by all these researchers are similar in both spirit and content, I'll focus on the one put forth by Omnés as representative of the overall class of proposals, which collectively I'll call the *consistent histories* approach to quantum reality.

The view of quantum mechanics presented by Omnés is that particles are definitely there and they have properties—but the properties are typically neither true nor false. As he puts it, "One must distinguish between the facts, the microscopic properties that may be said to be true, and also the vast number of microscopic properties that cannot even be said to be true or false." To illustrate this point, consider the property "the atom is green." This is a property that cannot be said to be either true or false, since an atom is much smaller than the wavelength of visible light. So it is meaningless to speak of the color of an atom. But it's reasonable to talk about the property that the atom is located in a certain region of space. In Omnés's view of the world, such meaningful microscopic properties have a very unusual structure.

Traditional quantum theory is based on the notion that the state of the system as characterized by the wave function determines the probability that the property is true when an appropriate measurement is made. The paradoxes and problems arise when we try to combine properties, since quantum theory admits no natural way to talk about the probability that the electron is in a certain region *and* that its momentum is in a certain range, for instance. The conventional Copenhagen view says that the combined property is simply meaningless; you can measure either the position or the momentum—but not both.

The Omnés program of quantum interpretation can be stated in a set of rules. Here I paraphrase the actual rules in a somewhat simplified form, in order to make them a bit easier to digest for the nonmathematical reader. The first rule is this:

RULE 1: *The theory of an individual isolated physical system is entirely formulated in terms of a specific space H in which the wave function resides, and a specific set of operations on the elements of H, together with the mathematical notions associated with them.*

Here Omnés uses the word "entirely" in the strongest possible sense. It means that not only the dynamics of the system but also the logical structure of the theory and the language we use when applying the theory are expressed in terms of the space *H* and the operations upon it. This rule assumes that everything that might be said about the physical system be stated in terms of its mathematical framework; in short, there is nothing else to be said about the system outside its mathematical formulation.

RULE 2: *The dynamics of the system are given by the Schrödinger wave equation.*

RULE 3: *Composite systems are described by "gluing together" copies of the spaces H for the individual systems in an appropriate way.*

At this point, Omnés invokes what in the jargon of quantum theory are termed *consistent histories* for the dynamical behavior of a quantum system. These histories are introduced in an attempt to eliminate the "collapse" of the wave function that we discussed earlier in this chapter. The idea underlying the Omnés view is that one should either explain the relation of this new dynamics to the dynamics of the Schrödinger equation or eliminate it from the theory. Consistent histories are a way of doing the latter.

The history of a quantum system is simply a listing of the properties that are observed at a sequence of times. Suppose that E and F are two different properties associated with different times, s and t, respectively, where s occurs before t. For these properties to be *consistent*, the probability of E followed by F plus the probability of *not-E* followed by F equals the probability of F. In this case, the history, namely, the properties E and F, is termed consistent. With this notion in mind, we have

RULE 4: *Every description of a physical system should be expressed in terms of properties belonging to a common consistent logic.*

Such a logical framework is provided by the probabilities associated with consistent histories. In this setup, a property implies another property if the probability of the second property, given that the first property has been observed, equals one.

In this picture, the consistent histories have to first explain classical properties at the macroscale, things like color, texture, and the alive or dead property of Schrödinger's Cat. These are properties defined by requiring that the variables defining them, what are often termed the "collective" variables, belong to a region of space that is much larger than the atomic scale defined by Planck's constant, which is defined as the energy of a photon vibrating at one cycle per second (1 Hz). In systems for which these collective variables have no significant interaction with variables on the atomic scale, these classical properties are essentially consistent and obey familiar deterministic dynamics of the sort defined by Newton's equations of motion. In particular, in these cases we can speak about the usual trajectories of the particles of the system given by classical mechanics à la Newton. In the event that there is interaction between the collective variables and the atomic-level environment, decoherence eliminates interference effects and allows the collective system to be described by ordinary probability theory.

The classical properties of tables, chairs, and cats that arise from this theory are called *phenomena* by Omnés; they are described by probabilistic laws. Such phenomena are termed *facts* when they actually occur in reality. So degrees of aliveness of Schrödinger's Cat are phenomena, and when the box is opened to see the cat as alive we have a fact. The passage from phenomena to facts, what Omnés calls *actualization,* involves a final rule:

RULE 5: *Physical reality is unique. It evolves in time in such a way that, when actual facts arise from identical antecedents, they occur randomly and their probabilities are those given by the theory.*

The problem with this rule for some is that it doesn't give a clear view of the mathematical formulation of actualization. In fact, Omnés remarks:

> The inability of quantum mechanics to offer an explanation, a mechanism, or a cause for actualization is in some sense a mark of its achievement. This is because it would otherwise reduce reality to bare mathematics and would correspondingly suppress the existence of time.

Rule 5 is aimed at blocking an interpretation of quantum mechanics in which all histories having a nonzero probability have to be given a claim to physical reality. This, in effect, is the Many-Worlds Interpretation of Everett, which Omnés rejects.

So far, we've spoken about macroscopic objects. What does Omnés's theory have to say about the properties of electrons, photons, and other microscopic objects? Do they have definite properties that can be said to be true or false? This is where the theory of measurement comes into play.

The Omnés theory considers the microscopic system, together with a larger system that acts as a measuring device. A photon plus a phosphorescent screen for registering its position would constitute such a system. The classical properties of the measuring device are called the *data* of the overall system, while the corresponding properties of the atomic system are the *results*. To be specific, then, the pinpoint of light on the screen would be the data of the electron-screen system, while the actual position of the electron would be the result. Under some circumstances, data and results fit into consistent histories and coalesce, one into the other. This is what constitutes a *measurement*. So the truth of properties of systems at the atomic level can only be settled definitively—true or false—in a measurement situation; otherwise, the electron being at a particular position cannot be said to be either true or false.

With this picture of goings-on at the microscopic level, Omnés's theory starts looking a lot like the conventional Copenhagen Interpretation—but with a twist. In the Copenhagen view, the classical, macroscopic world is complementary to the quantum world, while in the Omnés picture it emerges from the quantum world. In the Copenhagen scheme of things, there is the external notion of measurement; in the Omnés account, the corresponding process is that spontaneous classical actualization of facts. But in both accounts, one has to rely on external systems to have the interpretation hang together.

There is much more to be said in a technical way about this interpretation of quantum reality. But I think this is about as far as I dare go in a general volume of this sort. The technically trained reader is urged to look at the references cited in the "To Dig Deeper" section for a fuller account of the Omnés view of the way the world is. Now let's try to summarize the evidence for and against an objective, observer-independent quantum reality.

THE APPEAL:
SUMMARY ARGUMENTS

Let's review the bidding. The question is whether objects—quantum or otherwise—possess definite properties at all times, independently of their being

observed/measured by an external system. The Prosecution says no way; the
Defense argues *au contraire*. The competing positions circa 1988 are outlined in Tables 7.1 and 7.2, and are described in detail in *Paradigms Lost*.
Here's a summary of the evidence accumulated over the past decade or so
lending support to one side or the other.

EVIDENCE	INVESTIGATOR(S)	EVIDENCE FAVORS
interaction-free measurements	Elitzur-Vaidman, Zeilinger	Defense
Zeno effect	Sudarshan-Misra, Wineland	Defense
quantum eraser	Englert	Defense
quantum chaos	Berry, Gutzwiler	Defense
reality of wave function	Aharonov-Anandan	Defense
nonlocality	Hardy-Jordan, Gisin	Defense
consistent histories/ decoherence	Omnés, Griffiths, Gell-Mann and Hartle, Zurek	Defense

Table 7.3. The evidence

THE DECISION: APPEAL UPHELD

As Table 7.3 so clearly indicates, every single piece of evidence tends to favor
the Defense's claim that there really is an objective reality "out there," completely independent of whether we're looking. Some of this evidence, like that
involving quantum chaos, only weakly favors the Defense, while other pieces
like the theories of decoherence and consistent histories are strong indicators
that the Moon really is there when no one's looking. But there's not a single
piece of new evidence pointing to the Prosecution's case for observer-
dependent realities.

As for how to bet among the competing Defense theories outlined in
Tables 7.2 and 7.3, earlier I cast my vote for the Many-Worlds picture of
Everett, mostly because it did away with the Quantum Measurement

Problem—but at the extravagant price of dispensing complete universes for all occasions. The consistent histories view propounded by Omnés et al. has this same overwhelmingly attractive feature—but not the extravagance. So in light of all current evidence, I endorse overturning the previous verdict and giving consistent histories primacy of position in the race for quantum reality.

To Dig Deeper

Chapter One. The Truth, the Whole Truth, and the Scientific Truth

DOWN AMONG THE SOCIOLOGISTS

Alan Sokal's already famous hoax was first published in

Sokal, A. "Transgressing the Boundaries—Toward a Transformative Hermeneutics of Quantum Gravity." *Social Text*, Spring/Summer 1996, 217–252.

An illuminating critique of the entire episode is given in

Weinberg, S. "Sokal's Hoax." *New York Review of Books*, 8 August 1996, 11–15.

Together with Jean Bricmont, Sokal later looked deeper into the postmodern intellectuals' abuse of science in the volume

Sokal, A., and J. Bricmont. *Fashionable Nonsense*. New York: Picador, 1998.

A critique of the Edinburgh school of sociology's take on scientific knowledge is given in

Gottfried, K., and K. Wilson. "Science as a Cultural Construct." *Nature*, 10 April 1997, 545–547.

Andrew Pickering's angry rebuttal to the attack leveled in the above article against his book, *Constructing Quarks*, was published in *Nature*, 5 June 1997, 543.

SCIENCE AND GOD

The survey in the text about scientists and their belief in God extended to a much broader set of questions than what is reported here. For a complete account see

Larson, E., and L. Witham. "Scientists Are Still Keeping the Faith." *Nature*, 3 April 1997, 435–436.

Other good sources for the ongoing seesaw between science and religion are

Easterbrook, G. "Science and God: A Warming Trend?" *Science*, 277 (15 August 1997), 890–893.

Margenau, H., and R. Varghese, eds. *Cosmos, Bios, Theos*. LaSalle, IL: Open Court, 1992.

THE SCIENTIFIC SCHEME OF THINGS
A short, popular account of the Titius-Bode law is given in
Matthews, R. "The Ghostly Hand That Spaced the Planets." *New Scientist,* 9 April
    1994, 13.

A more complete discussion of empirical relations, laws of nature, and scientific
    theories is given in the opening chapter of *Paradigms Lost.*

PORTRAITS OF THE SCIENTIST IN WORDS
Roslynn Haynes's account of the myths that writers love to employ to characterize
    scientists is found in
Haynes, R. "The Power of Myth." *Helix,* no. 2 (1996), 12–19.

Chapter Two. That's Life!

General References

Here are a few introductory volumes that give a first-rate account of the ideas and
    problems associated with the question of the origin of life:
Morowitz, H. *Beginnings of Cellular Life.* New Haven, CT: Yale University Press,
    1992.
Trân Thanh Vân, J. and K., et al., eds. *Frontiers of Life.* Gif-sur-Yvette, France: Édi-
    tions Frontières, 1992.
Eigen, M. *Steps Towards Life.* Oxford: Oxford University Press, 1992.
Deamer, D., and G. Fleischaker. *Origins of Life.* Boston: Jones and Bartlett, 1994.

This last volume is especially noteworthy, as it contains reprints of many of the pio-
    neering papers establishing the various positions in the field.

WHAT IS LIFE?
The bookshelves sag under the weight of volumes purporting to define "life" in one
    way or another. A representative sampling of this cornucopia includes
Dawkins, R. *The Blind Watchmaker.* New York: Norton, 1987.
Dyson, F. *Origins of Life.* Cambridge: Cambridge University Press, 1985.
Levy, S. *Artificial Life.* New York: Pantheon, 1992.
Kuppers, B.-O. *Information and the Origin of Life.* Cambridge, MA: MIT Press,
    1990.

Needless to say, no one has yet succeeded in offering a definition of life that meets
    with universal acclaim. Nevertheless, everyone seems to know life when they see
    it. There have also been several nice survey papers published, outlining in more
    detail some of the competing theories—and problems with theories—just
    sketched in the text.
Baltscheffsky, H., et al. "On the Origin and Evolution of Life: An Introduction."
    J. Theor. Biol., 187 (1997), 453–459.
Pirie, N. "The Origins of Life on Earth." *Interdisciplinary Science Reviews,* 19
    (1994), 13–21.

De Duve, C. "The Beginnings of Life on Earth." *American Scientist,* 83 (September–October 1995), 428–437.

Ruse, M. "The Origin of Life: Philosophical Perspectives." *J. Theor. Biol.,* 187 (1997), 473–482.

THREE HURDLES ON THE WAY TO A THEORY

Accessible introductions to the origin of the genetic code are found in the articles

Hayes, B. "The Invention of the Genetic Code." *American Scientist,* 86 (January–February 1998), 8–14.

Schmidt, K. "Proofreaders for the Code of Life." *New Scientist,* 6 August 1994, 23–26.

A deeper look at the utility of "junk" DNA is given in

Nowak, R. "Mining Treasures from Junk DNA." *Science,* 263 (4 February 1994), 608–610.

Wells, W. "Don't Write Off 'Junk DNA.' " *New Scientist,* 8 June 1996, 19.

Buchbinder, H. "Talking Trash." *The Sciences,* May/June 1995, 8–9.

A good account of the handedness of the solar system is found in

Chyba, C. "A Left-Handed Solar System?" *Nature,* 389 (18 September 1997), 234–235.

IN THE BEGINNING WAS . . . RNA

An overview of the RNA world scenario is painted in the review article

Joyce, G. "RNA Evolution and the Origins of Life." *Nature,* 338 (16 March 1989), 217–224.

More recent evidence favoring this road to life is found in

Schimmel, P., and R. Alexander. "All You Need Is RNA." *Science,* 281 (31 July 1998), 658–659.

Lewis, R. "Primordial Soup Researchers Gather at Watering Hole." *Science,* 277 (22 August 1997), 1034–1035.

Picirilli, J. "RNA Seeks Its Maker." *Nature,* 376 (17 August 1995), 548–549.

Day, S. "The First Gene on Earth." *New Scientist,* 8 November 1991, 36–40.

LIFE FROM THE STARS

A good introductory account of the Martian meteorite brouhaha is given in the article

Melosh, H. "Blast Off." *The Sciences,* July/August 1998, 40–46.

Life from the stars is discussed in many places. Some representative accounts include

Chown, M. "Seeds, Soup and the Meaning of Life." *New Scientist,* 17 August 1996, 6.

Parsons, P. "Dusting Off Panspermia." *Nature,* 383 (19 September 1996), 221–222.

Travis, J. "Hints of First Amino Acid Outside Solar System." *Science,* 264 (17 June 1994), 1668.

Pendleton, Y., and D. Cruikshank. "Life from the Stars?" *Sky and Telescope,* March 1994, 36–42.

Talbot, R. "The Influence of the Interstellar Medium on Climate and Life." *Interdisciplinary Science Reviews,* 5 (1980), 102–111.

As a source for organic molecules, the asteroid impact theories hold considerable promise. A good survey of these results is given in

Chyba, C., and C. Sagan. "Endogeneous Production, Exogeneous Delivery and Impact-Shock Synthesis of Organic Molecules: An Inventory for the Origins of Life." *Nature,* 355 (9 January 1992), 125–132.

MODELING CLAY—AND LIFE

The mineral basis for the origin of life is discussed in

Von Kiedrowski, G. "Primordial Soup or Crêpes?" *Nature,* 381 (2 May 1996), 20–21.

Cairns-Smith, G. "The Origin of Life: Clays." *Frontiers in Biology.* Vol. 1, *From Atoms to Mind,* W. Gilbert and G. Valentini, eds. Milan, Italy, in press.

Cairns-Smith, G. "The Chemistry of Materials for Artificial Darwinian Systems." *Intl. Rev. in Phys. Chem.,* 7 (1988), 209–250.

SOME LIKED IT HOT

A popular treatment of life at the thermal vents is given in

Day, S. "Hot Bacteria and Other Ancestors." *New Scientist,* 9 April 1994, 21–25.

The iron sulfide theory of how life got going is given an airing in

Williams, R. "Iron and the Origin of Life." *Nature,* 343 (18 January 1990), 213–214.

SELF-REPLICATING LONG SHOTS

A brief introduction to Ghadiri's self-replicating proteins is given in

Cohen, P. "Are Proteins Real Key to Life?" *New Scientist,* 10 August 1996, 16.

Morowitz's membrane theory is outlined in great detail in his book cited under General References above.

THE STRUGGLE FOR SOULS

A detailed account of the creationist movement is provided in the volume

Numbers, R. *The Creationists: The Evolution of Scientific Creationism.* Berkeley, CA: University of California Press, 1992.

As for the creationists' latest attempts to get equal time in the classrooms, see

Scott, E. "Monkey Business." *The Sciences,* January/February 1996, 20–25.

Schmidt, K. "Creationists Evolve New Strategy." *Science,* 273 (26 July 1996), 420–422.

Jones, R. "Evolution and Creationism: The Consequences of an Analysis for Education." *Interdisciplinary Science Reviews,* 12 (1987), 324–332.

Chapter Three. Genetic Imperialism

General References

First-rate accounts of the general issues constituting the sociobiology debate are given in the "To Dig Deeper" section of *Paradigms Lost.* So let me repeat just a couple of the best here, starting with Edward O. Wilson's classic volume that sparked off the debate.

Wilson, E. O. *Sociobiology: The New Synthesis.* Cambridge, MA: Harvard University Press, 1975.

Other notable items, each of which shed its own particular light on the subject, include
Kitcher, P. *Vaulting Ambition.* Cambridge, MA: MIT Press, 1985.
Ruse, M. *Sociobiology: Sense or Nonsense?* Dordrecht, Netherlands: Reidel, 1979.
Caplan, A., ed. *The Sociobiology Debate.* New York: Harper & Row, 1978.

IT'S IN THE GENES
Oxford biologist Richard Dawkins has certainly been among the most vocal and visible proponents of the "genes-first" school of human behavior. A very informative and enlightening account of Dawkins's position in the sociobiological debate is found in
Stove, D. "The Demons and Dr. Dawkins." *American Scholar,* 61:1 (1992), 67–78.

For a good discussion of how genes influence the execution of complex behaviors in courtship and mating in fruit flies, see
Greenspan, R. "Understanding the Genetic Construction of Behavior." *Scientific American,* April 1995, 72–78.

The Swedish-American study on IQ levels in twins is reported in
McClearn, G., et al. "Substantial Genetic Influence on Cognitive Abilities in Twins 80 or More Years Old." *Science,* 276 (6 June 1997), 1560–1563.

NURTURE KNOWS BEST
The macaque monkey story is discussed in
Normile, D. "Habitat Seen Playing Larger Role in Shaping Behavior." *Science,* 279 (6 March 1998), 1454–1455.

In addition to the original account of memes as presented in the final chapter of Richard Dawkins's classic work, *The Selfish Gene,* the reader might want to consult the following volume for other work in the same direction:
Lynch, A. *Thought Contagion.* New York: Basic Books, 1996.

Evolution and religion have always been uneasy companions. An interesting interchange of letters on the topic appeared in the pages of *Nature* a few years back. The citations are
"Religion in the Genes." *Nature,* 362 (15 April 1993), 583.
"Evolution and Religion." *Nature,* 366 (25 November 1993), 296.

See also the popular article
Ridley, M. "Infected with Science." *New Scientist,* 25 December/1 January 1994, 22–24.

Lamarckian inheritance is about as popular with evolutionary biologists as final causation is with physicists. But the idea never seems to die. Papers arguing the pros and cons of the Cairns et al. work on bacteria are
Cairns, J., J. Overbaugh, and S. Miller. "The Origin of Mutants." *Nature,* 335 (8 September 1988), 142–145.

Lenski, R., and J. Mittler. "The Directed Mutation Controversy and Neo-Darwinism." *Science,* 259 (8 January 1993), 188–194.

THE SELFISH ALTRUIST

Altruism in all its many forms has formed the bedrock upon which much of the argument in favor of sociobiology rests. Different forms of altruism and how they arise in the animal world are discussed in

Clutton-Brock, T., and G. Parker. "Punishment in Animal Societies." *Nature,* 373 (19 January 1995), 209–216.

The theoretical work on indirect reciprocity by Sigmund and Nowak is reported in

Nowak, M., and K. Sigmund. "Evolution of Reciprocity by Image Scoring." *Nature,* 393 (11 June 1998), 573–577.

The question of what unit—gene, individual, group—natural selection acts upon to decide which organisms pass into the next generation has always been a contentious one. Classical Darwinism says it's the individual phenotype, Dawkins turned this upside down by claiming it is the gene, and there has always been a vocal minority saying they are both wrong, that it is an entire population. Pros and cons of this latter "group selection" hypothesis are discussed in

Wilson, D., and E. Sober. "Re-Introducing Group Selection to the Human Behavioral Sciences." *Behavioral and Brain Sciences,* 1994, 1–39.

A popular piece scouting out much of the same territory is

Bower, B. "Return of the Group." *Science News,* 148 (18 November 1995), 328–330.

IT'S ALL IN THE GAME

The original work setting up the theoretical basis for the use of game-theory ideas in biology and ecology is

Maynard Smith, J. *Evolution and the Theory of Games.* Cambridge: Cambridge University Press, 1982.

The work described in the text can be found in the articles

Mesterton-Gibbons, M., and E. Adams. "Animal Contests as Evolutionary Games." *American Scientist,* 86 (July–August 1998), 334–341.

Pool, R. "Putting Game Theory to the Test." *Science,* 267 (17 March 1995), 1591–1593.

PLAY IT AGAIN, SAM

Almost assuredly, the Prisoner's Dilemma game and its many offshoots constitute the most well-studied conflict resolution situation ever developed. Theoretical discussions of this can be found almost everywhere, including

Rapoport, A., and A. Chammah. *Prisoner's Dilemma: A Study in Conflict and Cooperation.* Ann Arbor, MI: University of Michigan Press, 1965.

Fascinating computer tournaments, in which different strategies are pitted against each other for the Prisoner's Dilemma game in a round-robin tournament, are reported in

Axelrod, R. *The Evolution of Cooperation.* New York: Basic, 1984.

Hofstadter, D. "Computer Tournaments of the Prisoner's Dilemma," in *Metamagical Themas,* pp. 715–734. New York: Basic, 1985.

Various extensions of the Prisoner's Dilemma game to bring it into closer contact with the real world are reported in the article
Axelrod, R. "Laws of Life." *The Sciences,* 27:2 (1987), 44–51.

Chapter Four. Born to Speak

General References

Excellent accounts of the language acquisition problem are given in the following volumes
Davis, J. *Mother Tongue.* New York: Birch Lane Press, 1994.
Pinker, S. *The Language Instinct.* New York: Morrow, 1994.
McCrone, J. *The Ape That Spoke.* New York: Morrow, 1991.
Bloom, P., ed. *Language Acquisition: Core Readings.* Cambridge, MA: MIT Press, 1993.

The modern approach to the study of language as a cognitive activity rather than an exercise in classification of types was introduced by Ferdinand Saussure in the early part of this century and was encapsulated in his famous book, *Cours de linguistique générale,* first published in 1916. An English translation is available in
Saussure, F. *Course in General Linguistics.* R. Harris, trans. La Salle, IL: Open Court Press, 1983.

The translator has given a lucid commentary on this seminal work in the companion volume
Harris, R. *Reading Saussure.* La Salle, IL: Open Court Press, 1987.

The reader should also consult the "To Dig Deeper" section in *Paradigms Lost* for many more sources of information on the problem of language acquisition.

THE LINGUISTIC WARS
The title of this section is taken from the following immensely informative and entertaining work, which outlines the rise and fall (and rise again) of the Chomskyan revolution:
Harris, R. *The Linguistic Wars.* New York: Oxford University Press, 1993.

A detailed account of the sad case of Genie is presented in
Rymer, R. "A Silent Childhood." *New Yorker,* 13 April 1992, 41–81, and 20 April 1992, 43–77.

A detailed account of the main arguments for and against the innateness hypothesis and the hypothesis of general learning is available in Chapter 4 of
Casti, J. *Paradigms Lost.* New York: Morrow, 1989.

ON HUMAN COMMUNICATION

The work by Petitto is well chronicled for the general reader in the Davis volume noted above.

For pointers to the animal language and origin of language literature, see the "To Dig Deeper" section of *Paradigms Lost.*

Robin Dunbar's work on grooming as the driving force in the development of language is outlined in his book
Dunbar, R. *Grooming, Gossip and the Evolution of Language.* London: Faber and Faber, 1996.

OUT OF THE MINDS OF BABES

John Locke's work on linguistic development in infants is recounted in his article
Locke, J. "Phases in the Child's Development of Language." *American Scientist,* 82 (September–October 1994), 436–445.

A small survey of recent work on the formation of the past tense of regular and irregular verbs is found in
Pinker, S. "Words and Rules in the Human Brain." *Nature,* 387 (5 June 1997), 547–548.

IT'S ALL IN THE BRAIN

A layman's account of the linguistic virtuosity of Christopher is given in
Blakeslee, S. "Brain Yields New Clues on Its Organization for Language." *New York Times,* 10 September 1991, B5–B6.

Petitto's work on the genetics of language is discussed in detail in the Davis book cited above under General References.

The work on Williams syndrome and its importance for language acquisition studies is discussed in
Fitzgerald, K. "Talking Genes." *The Sciences,* May–June 1992, 7–8.

CHOMPING ON CHOMSKY

Geoffrey Sampson's "devastating" critique of the Chomsky argument for language acquisition is laid out in
Sampson, G. "Language Acquisition: Growth or Learning?" *Philosophical Papers,* 18:3 (1989), 203–240.

"Learnability" as a central criterion for the evolutionary development of language is pursued in
Johansson, C. *A View from Language.* Travaux de l'Institut de Linguistique de Lund 34, Lund, Sweden: Lund University Press, 1997.

TOO HARD TO HANDLE

A gripping account of the decipherment of *rongorongo* is found in Steven Fischer's book
Fischer, S. *Glyphbreaker.* New York: Copernicus Books, 1997.

A brief discussion of the computer work used in deciphering the script on the St. Peter's cross is found in

Geake, E. "Aethelburg Knew . . ." *New Scientist,* 9 April 1994, 9.

Chapter Five. Man-Made Minds

General References

Excellent accounts of the philosophical aspects of artificial intelligence that this chapter focuses upon are found in the volumes

Copeland, J. *Artificial Intelligence: A Philosophical Perspective.* Oxford: Blackwell, 1993.

Moody, T. *Philosophy and Artificial Intelligence.* Englewood Cliffs, NJ: Prentice-Hall, 1993.

Dietrich, E., ed. *Thinking Computers and Virtual Persons.* San Diego, CA: Academic Press, 1994.

The journal *Artificial Intelligence* is a kind of "watering hole" where workers in the field gather to exchange ideas and opinions. One of the liveliest parts of this journal is the Book Review section, which traditionally allows a great deal of range and flexibility in what reviewers can say about the volumes under review. Many of the best reviews have been gathered together in the following work, which thereby constitutes a first-rate overview of the entire field:

Clancey, W., S. Smoliar, and M. Stefik, eds. *Contemplating Minds.* Cambridge, MA: MIT Press, 1994.

Another one-stop source for a glimpse of the whole field of AI is in the following volume, which consists of reprints of thirty of the most important papers in the history of AI, including Turing's original 1950 paper that sparked off the field:

Luger, G., ed. *Computation and Intelligence.* Cambridge, MA: MIT Press, 1995.

For a discussion of the problems and accomplishments of contemporary AI from the mouths of the researchers themselves, the collection of interviews with people like Allen Newell, John Searle, Herbert Simon, and David Rumelhart is not to be missed. These interviews are found in

Baumgartner, P., and S. Payr, eds. *Speaking Minds.* Princeton, NJ: Princeton University Press, 1995.

A good, but slightly technical, overview of the various *specific* approaches to AI is found in

Nilsson, N., and D. Rumelhart. "Approaches to Artificial Intelligence." Working Paper 93–08–052, Santa Fe Institute, 1993.

SLAUGHTER ON SEVENTH AVENUE

The problem of computer game-playing has fascinated researchers from the very inception of the field. Mostly, this is because games constitute a microworld with very clear-cut boundaries. Moreover, for the most part successful game-playing involves solving logical puzzles of the type a machine is good at. Here is a selec-

tion of books and articles that give a good account of how computers do it—and
how well they do it:

Bell, A. *The Machine Plays Chess*. Oxford: Pergamon Press, 1978.

Newborn, M. *Kasparov Versus Deep Blue: Computer Chess Comes of Age*. New York:
Springer, 1997.

Schaeffer, J. *One Jump Ahead: Challenging Human Supremacy in Checkers*. New
York: Springer, 1997.

Geake, E. "Playing to Win." *New Scientist,* 19 September 1992, 24–25.

Mechner, D. "All Systems Go." *The Sciences*, January/February 1998, 32–37.

The Turing test, the Chinese Room, and Gödel's theorem have all been well chronicled with many pointers to the literature in numerous places, including the "To Dig
Deeper" section of *Paradigms Lost*. So we will not repeat these references here.

SCHOOLS FOR MACHINES

The general approach to AI advocated by top-down workers is covered in many of
the volumes cited above under General References. Ditto for bottom-up.

THE CREATIVE COMPUTER

Accounts of the Loebner competition are given in the 5 November and 9 November
1991 issues of *The New York Times*. See also the article

Flood, G. "If Only They Could Think." *New Scientist,* 13 January 1996, 32–35.

Sources for more information on the expert systems, *Racter* and *BACON,* respectively, are

The Policeman's Beard Is Half-Constructed. New York: Warner Books, 1984.

Simon, H. "Computer Modeling of Scientific and Mathematical Discovery
Processes." *Bulletin of the American Mathematical Society*, 11 (1984), 247–262.

For a discussion of why programs like *Automated Mathematician* and *Eurisko* work, see

Lenat, D., and J. S. Brown. "Why *AM* and *Eurisko* Appear to Work," *Artificial
Intelligence,* 23 (1984), 269–294.

The *Cyc* project is described in the popular article

Davidson, C. "Common Sense and the Computer." *New Scientist,* 2 April 1994,
30–33.

MATTER MATTERS

An excellent recent text outlining the state of play in neural computing is

Mehrotra, K., C. Mohan, and S. Ranka. *Elements of Artificial Neural Networks*.
Cambridge, MA: MIT Press, 1997.

Another fine introductory account of neural networks is the volume

Aleksander, I., and H. Morton. *An Introduction to Neural Computing*. London:
Chapman and Hall, 1990.

One of the first "name-brand" scientists to lend public support to work on neural
networks was Nobel Prize winner Francis Crick. His call to arms is available in
the article

Crick, F. "The Recent Excitement About Neural Networks." *Nature,* 337 (1989), 129–132.

DID YOU SAY CHINESE?

The most recent of John Searle's seemingly countless attempts to explain the "obvious" merit of the Chinese Room argument is found in

Searle, J. *The Mystery of Consciousness.* New York: New York Review of Books, 1997.

GÖDEL'S NEW CLOTHES

Roger Penrose's broadside against strong AI, together with a follow-up account "explaining" what he meant, is available in the works

Penrose, R. *The Emperor's New Mind.* Oxford: Oxford University Press, 1989.

Penrose, R. *Shadows of the Mind.* Oxford: Oxford University Press, 1994.

The first of these volumes was extensively reviewed—and its arguments reviled! One such review, by John McCarthy, served as the source of the small dialogue between Penrose and a computer repeated in the text. Some of these reviews can be found in

McCarthy, J. Review of *The Emperor's New Mind. Bulletin of the American Mathematical Society,* 23 (1990), 606–616.

Landauer, R. "Is the Mind More Than an Analytical Machine?" *Physics Today,* June 1990, 73–75.

Johnson, G. "New Mind, No Clothes." *The Sciences,* July–August 1990, 45–49.

Dayan, P., and G. Wilson. Review of *The Emperor's New Mind. Network,* 1 (1990), 127–131.

Maynard Smith, J. "What Can't the Computer Do?" *New York Review of Books,* 37 (15 March 1990), 21–25.

Hays, D. Review of *The Emperor's New Mind. Journal of Social and Biological Structures,* 13 (1990), 179–183.

Need I say that not a single one of these reviews accepted any part of Penrose's arguments?

Chapter Six. Who Goes There?

General References

The SETI literature is just plain huge. Many first-rate volumes for the general reader are cited in *Paradigms Lost,* so I'll content myself with listing just a few new additions since the time of that book. The first item is a general overview of the SETI problem that is compact and covers all the important issues discussed here.

Ashpole, E. *The Search for Extraterrestrial Intelligence.* London: Blandford Press, 1989.

It's always interesting to hear what the pioneers of any field were thinking at the time they opened up a new area of intellectual endeavor. Interviews with the founders of the SETI movement are presented in

Swift, D., ed. *SETI Pioneers.* Tucson, AZ: University of Arizona Press, 1990.

Frank Drake initiated the first radio search for ETI in 1960 with Project Ozma. His autobiography tells a lot about not only that search, but also the development of the Drake equation.

Drake, F., and D. Sobel. *Is Anyone Out There?* New York: Delacorte Press, 1992.

An all-star cast of contributors to every aspect of the SETI question is found in

Zuckerman, B., and M. Hart, eds. *Extraterrestrials—Where Are They?* Cambridge: Cambridge University Press, 1995.

Finally, there is the always fascinating question of how the psychology of us Earthlings interfaces with the SETI issue. In what way does the enormous interest in alien contact on the part of the public stem from deep-seated psychological factors buried in the human mind? For some answers, see

Hough, P., and J. Randles. *Looking for the Aliens: A Psychological, Scientific and Imaginative Investigation.* London: Blandford Press, 1991.

A PLETHORA OF PLANETS

The search for and discovery of extrasolar planets is reported in

Mammana, D., and D. McCarthy. *Other Suns. Other Worlds?* New York: St. Martin's Press, 1995.

Halpern, P. *The Quest for Alien Planets.* New York: Plenum Press, 1997.

An account of simulations aimed at creating extrasolar planetary systems is summarized in

Gladman, B. "Twenty-eight Ways to Build a Solar System." *Nature,* 396 (1998), 513–514.

See also the articles

Glanz, J. "Worlds Around Other Stars Shake Planet Birth Theory." *Science,* 276 (1997), 1336–1339.

Mayor, M., and D. Queloz. "A Jupiter-Mass Companion to a Solar-Type Star." *Nature,* 378 (1995), 355–359.

WHAT WOULD HAPPEN IF THE TAPE WERE RUN TWICE?

Designing possible alien life-forms and speculating on their behavioral patterns is always a fun exercise. Here are a couple of accounts in that direction:

Lionni, L. *Parallel Botany.* New York: Knopf, 1977.

Cohen, J. "How to Design an Alien." *New Scientist,* 21/28 December 1991, 18–21.

Coffey, E. J. "The Improbability of Behavioral Convergence in Aliens—Behavioral Implications of Morphology." *Journal of the British Interplanetary Society,* 38 (1985), 515–520.

THE MIND OF AN ALIEN

The problem of cognitive universals is well covered in the article

Narens, L. "Surmising Cognitive Universals for Extraterrestrial Intelligences," in *Astronomical and Biochemical Origins and the Search for Life in the Universe,* C. Cosmovici et al., eds. Bologna, Italy: Editrice Compositori, 1997.

The question of earthly intelligences in other species like dolphins and whales is considered in

Falk, D. "Brain Evolution in Dolphins, Humans and Other Mammals: Implications for ETI." *Progress in the Search for Extraterrestrial Life,* ASP Conference Series, G. Shostak, ed., vol. 74, 1995, 53–62.

TALKING THE TALK
Papers exploring the issue of communication with ETI by language include

De Vito, C. "Languages, Science and the Search for Extraterrestrial Intelligence." *Interdisciplinary Science Reviews,* 16: 1 (1991), 156–160.

Augenstein, B. "Interstellar Communication by Visiting Card." *Journal of the British Interplanetary Society,* 43 (1990), 235–240.

Schmidt, S. "*Really* Alien Languages." *Analog Science Fiction and Fact,* July 1993, 4–12.

The deliberations and conclusions of the Department of Energy study to mark the Waste Isolation Pilot Plant are given in

The Development of Markers to Deter Inadvertent Human Intrusion into the Waste Isolation Pilot Plant—A and B Team Reports. Albuquerque, NM: Sandia National Laboratories, 22 April 1992.

ACCORDING TO GOTT
J. Richard Gott III explains his Principle of Indifference in

Gott, J. "Implications of the Copernican Principle for Our Future Prospects." *Nature,* 363 (1993), 315–319.

Gott, J. "A Grim Reckoning." *New Scientist,* 15 November 1997, 36–39.

Some rebuttals to Gott's arguments are found in the Letters to the Editor column of the 10 March 1994 issue of *Nature.*

THE SEARCH GOES ON
A full account of the Galileo flyby experiment to detect intelligent life on Earth is found in

Sagan, C., et al. "A Search for Life on Earth from the Galileo Spacecraft." *Nature,* 365 (1993), 715–721.

Those interested in building their own SETI listening station can get further information from the web site http://seti1.setileague.org/homepg.htm.

CONTACT!
The discussion of the characteristics of an ETI signal follows

Tarter, D. "Interpreting and Reporting on an ETI Discovery." *Space Policy,* May 1992, 137–148.

Some of the possible legal and political implications of contact are discussed in

Fasan, E. "Discovery of ETI: Terrestrial and Extraterrestrial Legal Implications." *Acta Astronautica,* 21 (1990), 131–135.

Goodman, A. "Diplomatic and Political Problems Affecting the Formulation and Implementation of an International Protocol for Activities Following the Detection of a Signal from Extraterrestrial Intelligence." *Acta Astronautica,* 21 (1990), 103–108.

See also

Billingham, J., et al., eds. *Social Implications of Detecting an Extraterrestrial Civilization.* Mountain View, CA: SETI Institute, March 1994.

Harrison, A. *After Contact: The Human Response to Extraterrestrial Life.* New York: Plenum Press, 1997.

Bova, B., and B. Preiss, eds. *First Contact.* London: Headline Books, 1990.

Chapter Seven. The Way the World Isn't

General References

Excellent accounts of the philosophical and physical aspects of quantum theory that this chapter focuses upon are given in

Wick, D. *The Infamous Boundary: Seven Decades of Controversy in Quantum Physics.* Boston: Birkhäuser, 1995.

Albert, D. *Quantum Mechanics and Experience.* Cambridge, MA: Harvard University Press, 1992.

Lindley, D. *Where Does the Weirdness Go?* New York: Basic Books, 1996.

The double-slit experiment is discussed in just about every book on quantum theory. The account given here follows that in

Hey, T., and P. Walters. *The Quantum Universe.* Cambridge: Cambridge University Press, 1987.

The versions of the Quantum Measurement and the Quantum Interpretation problems given here, as well as a far more detailed account of the various competing positions in the quantum reality game, are given in Chapter 7 of

Casti, J. *Paradigms Lost.* New York: Morrow, 1989.

Bell's theorem and its myriad consequences for the way things work in the quantum realm are explored in the volume

Cushing, J., and E. McMullin, eds. *Philosophical Consequences of Quantum Theory: Reflections on Bell's Theorem.* Notre Dame, IN: University of Notre Dame Press, 1989.

DECOHERENCE AND THE CAT'S KITTENS

A good nontechnical source for information about decoherence is in

Gell-Mann, M. *The Quark and the Jaguar.* New York: Freeman, 1994.

Other good sources include

Omnés, R. "Consistent Interpretations of Quantum Mechanics." *Reviews of Modern Physics,* 64 (1992), 339–382.

Leibfried, D., T. Pfau, and C. Monroe. "Shadows and Mirrors: Reconstructing Quantum States of Atom Motion." *Physics Today*, April 1998, 22–28.

The sad tale of Schrödinger's Cat is told in absolutely every popular work on quantum theory—this volume being no exception. Ditto for the Correspondence Principle. So readers wishing a different approach to the Measurement Problem have lots of alternatives to choose among.

THE UNCERTAINTY OF UNCERTAINTY
The Elitzur and Vaidman method is described in the article
Kwiat, P., H. Weinfurter, and A. Zeilinger. "Quantum Seeing in the Dark." *Scientific American*, November 1996, 52–58.

This article also serves as an excellent account of the work of the Zeilinger group on interaction-free measurement.

The quantum Zeno effect is chronicled in
Knight, P. "Watching a Laser Hot-Pot." *Nature*, 344 (5 April 1990), 493–494.
Ohja, P. "A Watched Atom Never Decays." *New Scientist*, 10 March 1990, 34.
Greenland, P. "Quantum Procrastination." *Nature*, 387 (5 June 1997), 548–549.

The discussion of the development of the Heisenberg Uncertainty relation follows that given in
Popper, K. *Quantum Theory and the Schism in Physics*. London: Unwin Hyman, 1982.

Englert's work on the quantum eraser is explained in
von Baeyer, H. C. "The Quantum Eraser." *The Sciences*, January/February 1997, 12–14.

The intriguing relationship between the classical dynamics of quantum theory as embodied in the Schrödinger equation and the classical phenomenon of chaos is discussed in
Monteiro, T. "Missing Chaos Challenges Rule of Quantum Mechanics." *New Scientist*, 29 June 1991, 25.
Jensen, R. "Bringing Order out of Chaos." *Nature*, 355 (13 February 1992), 591–592.
Brown, J. "Where Two Worlds Meet." *New Scientist*, 18 May 1996, 26–30.

WAVICLES
The Bohm–de Broglie pilot-wave theory is well expounded in the works
Holland, P. *The Quantum Theory of Motion*. Cambridge: Cambridge University Press, 1993.
Bohm, D., and B. Hiley. *The Undivided Universe*. London: Routledge, 1993.
Ben-Dov, Y. "De Broglie's Causal Interpretations of Quantum Mechanics." *Annales de la Fondation Louis de Broglie*, 14 (1989), 343–360.
Albert, D. "Bohm's Alternative to Quantum Mechanics." *Scientific American*, May 1994, 58–67.

Horgan, J. "Last Words of a Quantum Heretic." *New Scientist,* 27 February 1993, 38–42.

For a discussion of the reality of the quantum wave function, see
Freedman, D. "Theorists to the Quantum Mechanical Wave: 'Get Real.' " *Science,* 259 (12 March 1993), 1542–1543.

TO BE OR NOT TO BE—THAT IS NOT THE QUESTION
The EPR thought experiment is chronicled in so many places that it's impossible to avoid. Any of the general references cited at the beginning of this section are good places to find it. So are the articles

Branning, D. "Does Nature Violate Local Realism?" *American Scientist,* 85 (March/April 1997), 160–167.
Dewdney, C., et al. "Spin and Non-Locality in Quantum Mechanics." *Nature,* 336 (8 December 1988), 536–544.
Gribbin, J. "The Man Who Proved Einstein Was Wrong." *New Scientist,* 24 November 1990, 43–45.

The Dutch-door account of the Hardy-Jordan Theorem follows the Branning article above.

TELEPORTATION, Q-BITS, AND COMPUTATION
Excellent introductory accounts of not only quantum computation but also other "nonstandard" models of computation are given in the volume

Calude, C., J. Casti, and M. Dinneen, eds. *Unconventional Models of Computation.* Singapore: Springer, 1998.

For material on how to use a quantum computer to "compute the uncomputable," see

Bennett, C. "Certainty from Uncertainty." *Nature,* 362 (22 April 1993), 694–695.
Ekert, A. "Shannon's Theorem Revisited." *Nature,* 367 (10 February 1994), 513–514.
Buchanan, M. "Beyond Reality." *New Scientist,* 14 March 1998, 27–30.

CONSISTENT HISTORIES
The consistent histories approach to the Quantum Interpretation Problem is recounted in great detail in the Omnés article cited earlier. This material is greatly expanded in the book

Omnés, R. *The Interpretation of Quantum Mechanics.* Princeton, NJ: Princeton University Press, 1994.

See also the excellent review of this work

Faris, W. Review of Roland Omnés, *The Interpretation of Quantum Mechanics. Notices of the American Mathematical Society,* 43 (November 1996), 1328–1339.

Index